PLANETARY RINGS
A Post-Equinox View

SECOND EDITION

Fully updated and expanded, this new edition presents a cutting-edge summary of planetary rings and a wealth of exciting space mission results that have greatly increased our understanding of rings in recent years, including Cassini's initial Saturn System mission and its extended Equinox and Solstice missions, and the New Horizons flyby of Jupiter.

Larry Esposito draws on his 40 years' research experience, including his position as a principal investigator within the Cassini mission team, to present all aspects of planetary rings, emphasizing ring history and evolution. The book introduces basic physical processes and simple mathematical approaches in a clear, accessible manner, including N-body and stochastic models of ring dynamics. Further revised chapters present highlighted topics including Saturn's F ring, Uranus' rings and moons, Neptune's partial rings, dusty rings, and Jupiter's ring–moon system, including information from Galileo and New Horizons. Cassini results are fully integrated throughout, including new images in color, and a helpful new Afterword links each of the ring images in the Cassini 'Hall of Fame' gallery to the relevant explanation in the text. The book is also supported by an online cache of selected images and videos from the NASA collection, making it easy to locate the most relevant and beautiful illustrative materials available.

This is a key resource for students studying planetary science, and a thorough overview for researchers and professionals in astronomy and space-mission research.

LARRY ESPOSITO is Professor of Astrophysical and Planetary Sciences at the University of Colorado, and also the principal investigator of the Ultraviolet Imaging Spectrograph (UVIS) experiment on the Cassini space mission to Saturn. He was chair of the Voyager Rings Working Group, and, as a member of the Pioneer Saturn imaging team, he discovered Saturn's F ring. Dr Esposito has participated in numerous US, Russian and European space missions and used the Hubble Space Telescope for its first observations of the planet Venus. His awards received include the Harold C. Urey Prize from the American Astronomical Society, the NASA Medal for Exceptional Scientific Achievement, and the Richtmyer Lecture Award from the American Association of Physics Teachers and the American Physical Society. Dr Esposito has served as chair of the National Academy of Sciences Committee on Planetary and Lunar Exploration (COMPLEX), and is a Fellow of the American Geophysical Union.

Cambridge Planetary Science

Series Editors: Fran Bagenal, David Jewitt, Carl Murray, Jim Bell, Ralph Lorenz, Francis Nimmo, Sara Russell

Books in the series

1. *Jupiter: The Planet, Satellites and Magnetosphere*[†]
   Edited by Bagenal, Dowling and McKinnon
   978-0-521-03545-3
2. *Meteorites: A Petrologic, Chemical and Isotopic Synthesis*[†]
   Hutchison
   978-0-521-03539-2
3. *The Origin of Chondrules and Chondrites*[†]
   Sears
   978-1-107-40285-0
4. *Planetary Rings*[†]
   Esposito
   978-1-107-40247-8
5. *The Geology of Mars: Evidence from Earth-Based Analogs*[†]
   Edited by Chapman
   978-0-521-20659-4
6. *The Surface of Mars*
   Carr
   978-0-521-87201-0
7. *Volcanism on Io: A Comparison with Earth*
   Davies
   978-0-521-85003-2
8. *Mars: An Introduction to its Interior, Surface and Atmosphere*
   Barlow
   978-0-521-85226-5
9. *The Martian Surface: Composition, Mineralogy and Physical Properties*
   Edited by Bell
   978-0-521-86698-9
10. *Planetary Crusts: Their Composition, Origin and Evolution*[†]
    Taylor and McLennan
    978-0-521-14201-4
11. *Planetary Tectonics*[†]
    Edited by Watters and Schultz
    978-0-521-74992-3

12. *Protoplanetary Dust: Astrophysical and Cosmochemical Perspectives*[†]
    Edited by Apai and Lauretta
    978-0-521-51772-0
13. *Planetary Surface Processes*
    Melosh
    978-0-521-51418-7
14. *Titan: Interior, Surface, Atmosphere and Space Environment*
    Edited by Müller-Wodarg, Griffith, Lellouch and Cravens
    978-0-521-19992-6
15. *Planetary Rings: A Post-Equinox View* (Second edition)
    Esposito
    978-1-107-02882-1

    [†]Reissued as a paperback

PLANETARY RINGS

A Post-Equinox View

SECOND EDITION

.

LARRY W. ESPOSITO

Laboratory for Atmospheric and Space Physics,
University of Colorado

CAMBRIDGE
UNIVERSITY PRESS

University Printing House, Cambridge CB2 8BS, United Kingdom

One Liberty Plaza, 20th Floor, New York, NY 10006, USA

477 Williamstown Road, Port Melbourne, VIC 3207, Australia

4843/24, 2nd Floor, Ansari Road, Daryaganj, Delhi - 110002, India

79 Anson Road, #06-04/06, Singapore 079906

Cambridge University Press is part of the University of Cambridge.

It furthers the University's mission by disseminating knowledge in the pursuit of education, learning and research at the highest international levels of excellence.

www.cambridge.org
Information on this title: www.cambridge.org/9781108447904

First edition © L. Esposito 2006
Second edition © Larry Esposito 2014

First published 2006
Second edition 2014
First paperback edition 2017

A catalogue record for this publication is available from the British Library

Library of Congress Cataloging in Publication data
Esposito, Larry, author.
Planetary rings : a post-equinox view / Larry Esposito. – Second edition.
pages   cm
Includes bibliographical references and index.
ISBN 978-1-107-02882-1 (hardback)
1. Planetary rings.   I. Title.
QB603.R55E875   2014
523.9′8 – dc23   2013044085

ISBN 978-1-107-02882-1 Hardback
ISBN 978-1-108-44790-4 Paperback

Additional resources for this publication at www.cambridge.org/esposito

To my wife, Diane, and daughters, Rhea and Ariel

Contents

Preface: A personal view of planetary rings   *page* xiii

1   Introduction: The allure of the ringed planets   1

2   Studies of planetary rings 1610–2013   6
    2.1   Overview of planetary ring exploration   11
    2.2   Summary   19

3   Diversity of planetary rings   20
    Summary   33

4   Individual ring particles and their collisions   35
    4.1   Collisions   35
    4.2   Ring particle size and shapes   39
    4.3   Dynamic ephemeral bodies or "rubble piles"   41
    4.4   The Roche limit   42
    4.5   Ring particle composition and temperature   44
    4.6   Are rings gases or liquids?   47
    4.7   Evolutionary rate   50
    4.8   Energy transfer   51
    4.9   Planetary rings as a gas: kinetic theory approach   51
    4.10 Planetary rings as a fluid: hydrodynamics   52
    4.11 Summary   53

5   Large-scale ring evolution   54
    5.1   Particle dynamics   54
    5.2   Mechanisms for ring confinement   55
    5.3   Summary   59

6   Moons confine and sculpt rings   60
    6.1   Resonances   60

	6.2	Pendulum model	61
	6.3	Shepherding	63
	6.4	Waves	66
	6.5	Longitudinal confinement	68
	6.6	Summary	69
7		Explaining ring phenomena	70
	7.1	Saturn's broad rings	70
	7.2	Waves	85
	7.3	Spokes	87
	7.4	Narrow and eccentric rings	90
	7.5	Dusty rings	92
	7.6	Clumpy rings	93
	7.7	Summary	100
8		*N*-body simulations	101
	8.1	Periodic boundary conditions	102
	8.2	Impacts	104
	8.3	Results of *N*-body simulations	104
	8.4	Stability, instability, and overstability	107
	8.5	Summary	110
9		Stochastic models	111
	9.1	Motivation	111
	9.2	Stochastic processes	112
	9.3	Random walk on the line	112
	9.4	Markov processes	112
	9.5	Markov chains and Monte Carlo simulations	113
	9.6	Stochastic processes as ring models	114
	9.7	Summary	114
10		Age and evolution of rings	116
	10.1	Introduction	116
	10.2	Age of the rings	117
	10.3	Age of the Jupiter rings	118
	10.4	Age of the Saturn rings	118
	10.5	Age of the Uranus rings	118
	10.6	Age of the Neptune rings	120
	10.7	Satellite disruption	121
	10.8	Ring formation	123
	10.9	Cosmic recycling	125
	10.10	Summary	134

11	Saturn's mysterious F ring	135
	11.1 F-ring structure	137
	11.2 F-ring shepherds	138
	11.3 Explanation for the shepherd wandering	143
	11.4 Cassini sees F-ring brightening	144
	11.5 Origin and history	147
	11.6 Summary	150
12	Uranus' rings and moons	151
	12.1 Description	151
	12.2 Dynamics and origin	152
	12.3 Summary	154
13	Neptune's partial rings	155
	13.1 Dynamical explanations	156
	13.2 Voyager observations	157
	13.3 Porco's model	157
	13.4 Problems with Porco's model	160
	13.5 Salo's model	161
	13.6 The deadly 1998 observations	162
	13.7 Namouni and Porco's alternative solution	166
	13.8 Keck observes the Neptune rings with adaptive optics	167
	13.9 Summary	170
14	Jupiter's ring–moon system after Galileo and New Horizons	172
	14.1 Cassini observations of Jupiter's rings	174
	14.2 Vertical corrugations	174
	14.3 Size distribution in Jupiter's main ring	174
	14.4 Summary	178
15	Ring photometry and spectroscopy	180
	15.1 Photometry	180
	15.2 Spectroscopy	184
	15.3 Summary	185
16	Dusty rings	187
	16.1 Forces on dust	187
	16.2 Dust lifetimes	191
	16.3 Dust clouds	192
	16.4 More dusty Jupiter rings	193
	16.5 Saturn's ethereal dusty rings	194
	16.6 Summary	202

17 Concluding remarks   203
     17.1 The big questions   206

     Afterword: The Cassini Hall of Fame   208
     Glossary   220
     References   226
     Index   241

Color plate section is between pages 130 and 131.

Preface: A personal view of planetary rings

I have spent most of my professional life trying to understand the rings around the giant planets. What do the observations tell us? What are the key phenomena, processes, and principles? What is the history of planetary rings?

My colleagues in planetary sciences and I have developed excellent, although still incomplete, answers to these important questions. These results are printed in a large number of journal articles, reviews, and books intended for graduate students and professionals. Unfortunately, we have no up-to-date books suitable for the beginning graduate student or interested undergraduate. It is my purpose to provide a summary of this broad area, including the latest results from the international Cassini mission. Cassini has been orbiting Saturn since 2004. In August 2009, the Sun set on the rings at the time of its vernal equinox. At that time, objects in the ring were spectacularly visible as they cast long shadows. Cassini provided images of small bodies embedded in Saturn's rings and spectacular views of vertical excursions at ring edges and other locations. Along with other Cassini observations, such as high-resolution measurements from star occultations and from spectroscopy of the dimly lit rings, these equinox findings have given a new perspective on planetary rings that is even more dynamic. Comets strike the rings; impacts destroy small moons; embedded transient objects appear. Particularly, the equinox observations may indicate that new objects are forming at the present time.

This book has several objectives. I have endeavored to produce a book that is accessible to the general reader; is suitable for a companion book to undergraduate astronomy electives and courses in general astronomy; and which can provide an entry point for research and reading by graduate students and professionals through an extensive bibliography. Although I cover all the major topics related to planetary rings, I admit an emphasis on some topics that have caught my own attention, particularly the origin of planetary rings and explaining the structure of the ring systems. The reader will indulge me where I discuss these examples in the latter part of the book.

The overall plan is as follows: all interested readers may read Chapters 1, 2, and 3, which provide a historical and descriptive overview, and a summary of the major results and big questions in Chapter 17. My discussion of the basic physics of the rings is covered in Chapters 4–7. This is accessible to undergraduate students in the physical sciences, with some small exceptions. A more in-depth treatment of modeling (Chapters 8 and 9), and some hot topics (Chapters 10–16) will reward the dedicated reader. Extensive references to the original papers allow a number of stepping-off points for term papers or deeper investigation. The most recent Cassini findings are highlighted in newly revised Chapters 7, 11 and 16.

I appreciate helpful discussions on the first edition with Glen Stewart, Tom Brophy, Robin Canup, Mihaly Horanyi, Mark Lewis, Joshua Colwell, Shawn Brooks, Jim Howard, Dimitri Veras, and Andre Brahic. Dick French read the entire draft manuscript of the first edition, and made many helpful suggestions. Phil Nicholson answered some particular questions. For the second edition, Doug Hamilton supplied the latest version of the pie chart in Chapter 3. Mihaly Horanyi and Matt Tiscareno reviewed several chapters each. I thank Jeff Cuzzi and Frank Spahn, who hosted my sabbatical visits while I worked on the second edition of this book. Laura Bloom again produced the final product. I thank the NASA Planetary Geology Program and Outer Planets Data Analysis Program, the Outer Planets Research Program, Jupiter System Data Analysis Program, and the Voyager, Galileo, and Cassini projects for supporting my research on planetary rings over the past 40 years. Of course, any errors are my own responsibility.

1

Introduction

The allure of the ringed planets

One of the most enduring symbols of space exploration is a planet surrounded by a ring. This symbol inspires a celestial context: nothing on Earth is like it. My own fascination with space began when first I saw Saturn and its rings through a small telescope. The study of planets and rings has become my lifelong vocation. What a wonderful surprise that the ringed planets are just as beautiful and scientifically compelling seen close up! Furthermore, the ringed planets are not just objects of beauty, but complicated physical systems that provide a local laboratory and analogy for other cosmic systems like galaxies and planet-forming disks.

We now know that planetary rings, once thought to be unique to the planet Saturn, exist around all the giant planets. These rings are not solid objects, but composed of countless particles with sizes from specks of dust to small moons. For each planet, the rings are quite different. Jupiter's ring is thin and composed of dust-like small particles. Saturn's rings are broad, bright, and opaque. Uranus has narrow, dark rings among broad lanes of dust that are invisible from Earth. Neptune's rings include incomplete arcs restricted to a small range of their circumference. All rings lie predominantly within their planet's Roche limit, where tidal forces would destroy a self-gravitating fluid body. They are also within the planet's magnetosphere, and in the case of Uranus, they are within the upper reaches of the planetary atmosphere.

The common occurrence of ring material around the planets is one of the major scientific findings of the last 40 years. The new ring systems were discovered by both spacecraft and ground-based observers, often surprising us by contradicting our expectations. The rings' appearance and composition differ among the various planets, and likewise within each ring system. The broadest set of rings and the most identified processes are found around the planet Saturn, which has been scrutinized by the US/European Cassini space mission since 2004.

In July 2004, the Cassini spacecraft began to orbit Saturn, with the mission planned to last four years. Just after the rockets fired to place Cassini into orbit, it took the closest, highest-resolution observations of Saturn's rings. From a

distance of within 20 000 km, Cassini cameras and spectrometers returned spectacular data. Carolyn Porco of the Space Sciences Institute and leader of the camera
team stated: "Our glistening golden-shrouded mechanical beast, with seven lonely
years and billions of miles behind it, set its sights on a softly-hued, giant and
dreamlike planet . . . We hasten soundlessly across a vast and solid sea of rippling
waves, scalloped shores and tumbling rivers of frozen boulders, stones, pebbles all
made of ice." This is her description of the first Cassini high-resolution views of
Saturn's rings. Cassini took close-up pictures of new structures and new moons
within the rings, measured the ring composition and its variation, saw evidence of
dynamical phenomena indicating the gravitational interactions with nearby moons,
and provided new findings about the ring evolution and history.

The detailed views from spacecraft, ground telescopes, and Hubble show a
variety of structural features in planetary rings: vertical thickness considerably
greater than the average particle size; dark lanes, gaps, and other density variations;
eccentric and inclined rings; sharp edges; azimuthal brightness variations, arcs, and
clumps; waves and wakes; and incomplete, kinked, and apparently braided rings.
We still lack good explanations for much of this dazzling variety of phenomena,
although many of these features have been explained by gravitational interactions
between the ring particles and nearby moons.

Beyond the interactions with moons (many of which were likewise discovered
by close-up pictures from spacecraft), the ring particles interact with the planet's
magnetosphere via charging, plasma drag, and forces from the planet's own magnetic and electric field. Electrostatic effects lift small particles from the surface of
the larger ring particles to create the dark radial lanes, called spokes, that are seen
in Voyager Saturn and Cassini pictures. Ring particles suffer a gas drag from the
extended planetary atmosphere that causes them to spiral inward to destruction.

Ring particles come in a broad range of sizes. Their size distribution extends
from submicron dust, through meter-sized particles, to small embedded moons like
Saturn's moons Pan and Daphnis, about 10 km in radius. Perhaps 100 to 1000 moons
bigger than 1 km orbit each of the giant planets but were too small to be detected
by spacecraft cameras. Theoretical expectations and some data support the idea
that the particles in a ring will segregate in size, both radially and vertically.

What are ring particles made of? The ring composition is well known only for
Saturn. Spectroscopic, occultation, and neutron measurements all imply that Saturn's rings' particles are almost entirely water ice. They are bright, like the surfaces
of Saturn's inner satellites. For the other ring systems, the particles resemble the
nearby small moons and probably contain significant amounts of silicate, and in the
case of Uranus and Neptune, possible carbonaceous material. Even in the Saturnian rings, color and spectral variations indicate compositional differences between

different parts of the rings. Some of these differences may be primordial; others arrive from interactions with the environment, including meteoroid bombardment.

Radio occultations at multiple wavelengths have provided size information for the Saturn and Uranus rings in the range of roughly 1 cm to 10 m. Information on smaller particles is from photometry and differential opacity in stellar occultations. Cassini's Cosmic Dust Analyzer has directly sampled the smallest ring particles. The broad ring particle size distributions are similar to those arising from catastrophic fragmentation of small solid bodies.

We have a first-order understanding of the dynamics and key processes in rings, much of it based on previous work in galactic and stellar dynamics. The rings are a kinetic system, where the deviations from perfect circular, equatorial motion can be considered as random velocities in a viscous fluid. Unfortunately, the models are often idealized (e.g., treating all particles as hard spheres of the same size) and cannot yet predict many phenomena in the detail observed by spacecraft (e.g., sharp edges). Collective effects can also give rise to unexpected structures. The latest Cassini data show spectacularly that the rings do not behave like a simple fluid.

The rings show many youthful features: Saturn's ice is bright and relatively undarkened by meteoritic dust, the Uranus rings are narrow, Neptune's arcs are constrained to a small range of longitude, and Jupiter's ring particles are so small that they will be dragged away into the planet's atmosphere in a thousand years or less. The angular momentum that is now being transferred between rings and the nearby moons through density waves should have caused them to spread much further apart than they are now. Further, the small moons discovered by Voyager could not themselves have survived the flux of interplanetary meteoroids for the age of the solar system. In much less time, these small moons would have been shattered by an impacting object. This realization provides a potential solution to the problem of young rings. These impacts not only destroy the moons; they can also recreate the ring systems. The new rings would gradually spread and eventually be ground to dust. Shattered moons can re-form, to provide material for future rings. Data from Cassini's observations at Saturn equinox indicate that the moons not only sculpt the rings' structure; they also provide the reservoirs for past and future ring systems, and possibly trigger new structures forming now.

The Voyager flybys of the 1980s changed forever how we think about planetary rings. Cuzzi (1998) describes this as a "paradigm shift" that led us to the current conception of young and dynamic rings. The new and exotic phenomena seen close up by the Voyager spacecraft forced us to expand our conceptions and adjust them to the immediate reality. Unfortunately, the Voyager investigations were spotty in coverage, both in time and space. The Cassini spacecraft benefits from the capability of experiments developed one generation later. And furthermore, from the long duration of its tour of the Saturn system, now extended to 13 years . . . , including the spectacular equinox, the moment when the Sun set on Saturn's rings in August 2009. Cassini's long mission allows us to address new questions and phenomena as they arise, and to follow up with deeper investigations. For example, some tantalizing bright clumps seen by Voyager in Saturn's F ring (Showalter, 1998) provided one example of targets for monitoring by Cassini's cameras. French *et al.* (2012) describe how the F ring is now spectacularly different from our Voyager view. Earlier predictions of small moons and self-gravity wakes within Saturn's rings have been confirmed by Cassini.

The Cassini capability exceeds that of Voyager in every area. The cameras resolve sub-kilometer structure in the rings. Multiday "movies" have tracked spokes, F-ring phenomena, and small satellites. More than 100 stellar and solar occultations define the radial and azimuthal structure in the rings, resolving features as small as a few meters. The Cassini experiments have observed the rings at wavelengths from the extreme ultraviolet to the mid-infrared. Radio occultations have defined the particle size distribution for particles with size from millimeters to tens of meters. The Cassini orbit provides a full range of observing geometry to catch the lit, unlit, shadowed, backlit, and full-phase (when the Sun is directly behind the spacecraft) rings. The combination of observing over this range of wavelengths and angles now gives us a much more complete picture of the vertical structure, particle temperatures, and particle shapes. Multiple observations have caught waves and wakes in action. The orbiter has measured the dust in the diffuse rings, around the rings, and in the bombarding meteorite population. The comprehensive Cassini payload allows us now to characterize the interactions of rings, atmosphere, iono-sphere, and magnetosphere as well as ring–satellite relations and ring evolution. Joseph Burns of Cornell University said that "Cassini results may fundamentally alter our understanding of ring processes and even origins." The recent results of many Cassini scientists confirm this expectation. Cassini has now produced more than 300 papers about Saturn's rings.

I will emphasize that most rings are much younger than the solar system and that the likely explanation is that new rings are episodically created by the destruction of small moons. This idea is one example of how recent spacecraft observations have shown a larger role for catastrophic events in the history of the solar system. The role of singular events, like giant impacts, seems essential in explaining the history of planetary rings, placing ring history now in a class with the origin of the Earth's moon, the history of life on Earth, the rotation and tilt of the planets, and unique aspects of both Mercury and Pluto. Thus, the study of rings connects with similar rare and sudden processes like the formation of the Earth's moon and the demise of the dinosaurs.

This unexpected range of phenomena seen in planetary rings gives some insight into the processes in other flattened astrophysical systems. The processes we now observe in planetary ring systems parallel those that occurred at the time of the origin of the planets. Clearly, the rings are not now accreting to form planets, as the original planetesimals did. However, many processes are occurring now in rings that resemble those in the solar nebula, particularly interactions between the disk and embedded protoplanets. Models that explain the present processes in rings can be compared in detail to ring observations, allowing testing and refinement that is no longer possible for the early solar system. In 2009, the Cassini space mission observed the Sun setting on the rings, which occurs every 15 years at Saturn's equinox. This unique viewing perspective allowed us briefly to see structures never before seen. Cassini continues to observe Saturn's rings until the mission's planned end in 2017, just after summer solstice on Saturn.

Fortunately for Cassini, the spacecraft has no plans to crash into the rings. Even small particles of 1 mm or so can be deadly, but they are likely to be rare outside the visible rings (Canup and Esposito, 1997). Unfortunately, this means that we will attain no close-up views of individual ring particles. NASA has considered plans for a future "Ring Observer" mission that would come close enough to the rings to hover and capture pictures of the individual particles. Such views would provide spectacular "ground truth" for the remote sensing from flyby and orbiter spacecraft.

This book briefly describes the rings in the solar system and some mathematical and physical approaches to understanding them. I will emphasize the variety of ring phenomena revealed in the last few decades from space and Earth observations. I address, in some detail, the recent history of rings and conclude with a discussion of the major open questions and future opportunities to learn about planetary rings. My presentation benefits and draws from previous reviews by Esposito *et al.* (1984), Esposito (1986), Esposito *et al.* (1991), Nicholson and Dones (1991), Esposito (1993), Porco *et al.* (1995), Cuzzi (1995, 1998), Burns *et al.* (2001), Esposito (2002), Dougherty *et al.* (2009), Cuzzi *et al.* (2010), Esposito (2010) and Tiscareno (2013b).

2

Studies of planetary rings 1610–2013

The possibility of rings surrounding the planets was not imagined by the ancients. Even the possibility that the planets possess moons was a surprise to Renaissance intellectuals. This latter finding was one of the first discoveries made by Galileo with the newly invented telescope in 1609, when he turned his view to Jupiter and discovered it to be accompanied by the four small objects Io, Europa, Ganymede, and Callisto. These bodies are now fittingly known as the Galilean satellites of Jupiter. The close-up investigations by space missions Pioneer, Voyager, Cassini, New Horizons and the Galileo space mission have revealed each satellite to be a unique world with its own history and geology. Galileo also turned his telescope to the Earth's moon, where he discovered its mountains; to Venus, where he discovered its phases; and to the Sun, where he discovered sunspots. Each of these new findings undercut the classical ideas of an Earth-centered and perfect celestial universe: they provided basic support for the Sun-centered Copernican model, where earthly laws applied universally.

In 1610, Galileo turned his telescope to Saturn. With his imperfect optics, it seemed that the planet had a giant moon to either side. But these "moons" were unlike the Jupiter satellites he had found previously: they apparently remained stationary. Furthermore, when he observed Saturn again in 1612, they had disappeared. Over the next decades, Saturn appeared in many different forms to the early telescopic observers. Their drawings give a bizarre array of inconsistent figures (see Figures 2.1 and 2.2) and further led to a variety of possible explanations (all incorrect) over the first half of the seventeenth century. Indeed, the nature of Saturn was one of the major astronomical controversies of that period. In 1656 the Dutch astronomer Christiaan Huygens deduced the correct explanation and published it as a Latin anagram, which when decoded and translated reads, "It [Saturn] is surrounded by a thin, flat ring, nowhere touching, and inclined to the ecliptic." The modern view of Saturn's seasonal changes can be seen in Figure 2.3. The flattened rings appear differently at each Saturn season, as viewed from the Earth; in

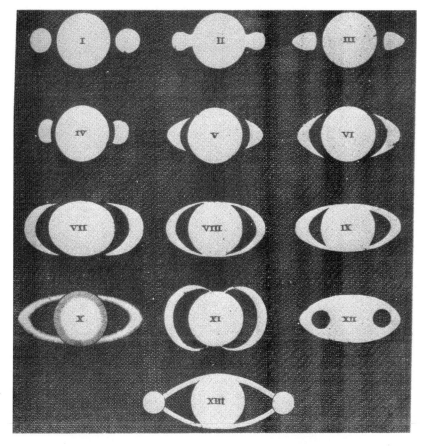

Figure 2.1 Seventeenth-century drawings of Saturn (from *Systema Saturnium*, 1659). I, Galileo (1610); II, Scheiner (1614); III, Riccioli (1641 or 1643); IV–VII, Hevel (theoretical forms); VIII, IX, Riccioli (1648–50); X, Divini (1646–48); XI, Fontana (1636); XII, Biancani (1616), Gassendi (1638–39); XIII, Fontana and others at Rome (1644–45). Riccioli made a drawing in 1646 rather like XI but less distorted. (Photographed from the copy of *Systema Saturnium*, © The Trustees of The British Museum.)

fact, they vanish when the Earth passes through the plane of Saturn's equator. This ring plane crossing phenomenon occurs about every 15 years, approximately at the equinoxes on Saturn. During 1995 to 1996, the Hubble Space Telescope observed the ring plane crossing events (see Figure 2.4). The main rings of Saturn are so thin that no light reflected from them at the exact moment of crossing has ever been observed, even by Hubble. Hubble saw mainly the reflection from Saturn's F ring, equivalent to a layer about 1.5 km thick (Nicholson *et al.*, 1996). Cassini close-up observations during the equinox of 2009 confirm that the rings are flat and thin,

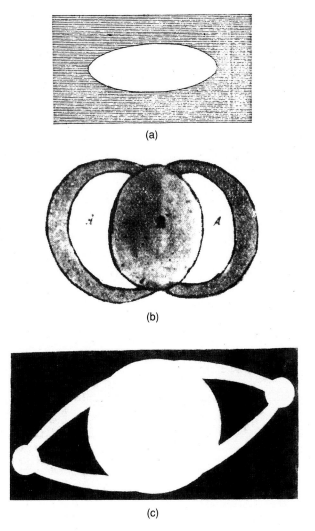

(a)

(b)

(c)

Figure 2.2 Drawings of observations by (a) Pierre Gassendi (1634), (b) Francesco Fontana (1638), and (c) Fontana (1645). Originals in *MSS Galileiana* 95, f. 81r, Biblioteca Nazionale Centrale, Florence, Italy. (By concession of the Ministero per i Beni e le Attività Culturali della Repubblica Italiana.)

but with occasional excursions up to a few kilometers in perturbed regions. (See Chapter 7.)

Huygens argued that his telescope was superior to those with whom he disputed, but this was not justified. In fact, his contemporaries had observed the same phenomena as he had but had not interpreted them as he did. The difference was not in seeing the rings, but in *perceiving* them. Huygens was a disciple of René Descartes, who developed the idea of a vortex as an explanation of motions in the

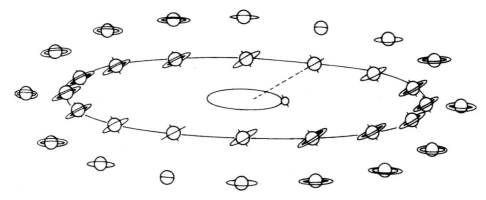

Figure 2.3 Cartoon views of Saturn and its rings over one Saturnian orbit according to Huygens' model (de Pater and Lissauer, 2010, after Huygens, 1656). The inner ellipse represents the Earth's orbit; the outer ellipse represents Saturn's orbit. The dashed line is the view from Earth. The outer drawings give the appearance of the rings as seen from Earth. (Reprinted with the permission of Cambridge University Press.)

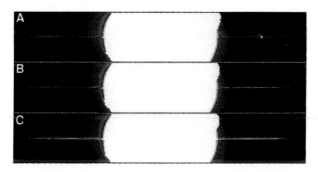

Figure 2.4 A sequence of Hubble images at 0.89 μm obtained on August 10, 1995 as the Earth crossed Saturn's ring plane, at (A) 20:12, (B) 21:49, and (C) 23:42 UT. Each frame is a composite of two images, processed to remove cosmic rays and small satellites. Dione is visible 17 arcsec west of Saturn in (A). The planet's disk is heavily saturated in these 300-s exposures. North is at the top, and east is to the left in all figures (from Nicholson *et al.*, 1996). (Reprinted with permission from *Science*, **272**, 509–15. Copyright 1996, American Association for the Advancement of Science.)

universe (van Helden, 1984). At the Creation, God had put all extended substance in the universe in motion in vortices. Our solar system was but one. Planets moved in the solar vortex and were themselves centers of smaller vortices. For example, Jupiter's vortex accounted for the motions of the Galilean satellites (Descartes, 1644, 80–202). Huygens thought in terms of *Cartesian* vortices. When he discovered Saturn's satellite Titan in 1655, he knew that Saturn's vortex extended that far, including the rings.

Since Huygens determined Titan's period to be 16 days, he believed that the rings' material must revolve faster, but not as fast as Saturn itself (which he concluded rotated in 1/2 day by analogy with the Earth and its moon). Since he saw no variations on this scale of days (the puzzling behavior that was observed extended over Saturn's orbital period of 29 years), he naturally thought in terms of a symmetric ring, flat and not varying, around its circumference. Only a body symmetrical around the axis of rotation of Saturn's vortex could solve this problem (van Helden, 1984). Thus, it is clear that Huygens was drawn to the correct solution as much (or more) by his philosophical conceptions as by his own observations. In the more recent history of planetary ring studies, a similar situation is evident: repeatedly we have found we need to overcome our preconceptions to understand the rings. Particularly, there is a need to revise the model of Saturn's rings that persisted up until the late 1970s as simple, circular, unchanging, equal-sized bodies orbiting the planet. Spacecraft and ground-based observations have forced us to drop the simple models and embrace a much more complicated and active view of planetary rings. The rings have continued to change between the Voyager and Cassini missions. Richard Greenberg and Andre Brahic (1984) note that this has forced the study of planetary rings to be dynamic and evolving: like the rings themselves! The observations continue to provide a sharp incentive, challenging us to extend our thinking and drop our preconceptions to truly understand the nature and history of planetary rings.

The interplay between observations and theory is clearly seen in the history of ring studies. From the first, the quality of the observations fell just short of revealing the rings' essential nature. Galileo's observations allowed multiple explanations. Huygens' revelation was the result of new thinking and application of the Cartesian world-view, rather than better observation. Similarly, Huygens imagined the ring around Saturn as a solid body. The discovery of a gap in Saturn's ring by Jean Dominique Cassini in 1676 (now known as the Cassini Division) led many to the idea that the rings were composed of multiple satellites. Nonetheless, Huygens never abandoned his original proposal of a solid, thin ring.

Saturn's ring served as a model for the overall structure of the universe in the cosmology of the seventeenth through nineteenth centuries. The philosopher Immanuel Kant applied this disk model to the "nebulae," or clouds, seen in the telescope, which we now know to be galaxies of countless stars. Kant proposed that God had brought about the current situation by using natural forces. Matter became concentrated in rotating nebulae and was flattened as a result of this rotation into individual galaxies, stars, and planets. This nebular hypothesis is still the paradigm for planet formation; and the rotation is presently seen as the explanation for the flatness of planetary rings. Thus, since the middle of the eighteenth century, the flat disk (of which Saturn's ring was the local example) has become an ever-present cosmic model.

The debate continued into the nineteenth century about the possible solidity of Saturn's rings. William Herschel believed Saturn was surrounded by two solid rings. Pierre Simon Laplace concluded that Saturn was surrounded by a large number of narrow solid rings. The idea of solid rings received a blow when Saturn's C ring was discovered in 1848, and it was found that the planet itself was visible through the ring. This was a motivation for its designation as the "crepe" ring, and later ring C. In 1857 James Clerk Maxwell won the Adams Prize for his classic essay that showed Saturn's rings could only consist of an indefinite number of small particles (Alexander, 1962; Brush *et al.*, 1983). Laplace and Maxwell had shown that a solid ring would be unstable. Later, James E. Keeler (1985) made the confirming observations by showing spectrographically that the orbital velocity of the constituent particles varies across the ring as predicted by *Kepler*'s laws for a collection of small satellites.

In the twentieth century, the extreme flatness of the rings was recognized and explained with dynamical models (Brahic, 1977; see Chapter 5). *Spectroscopy* showed water ice as the dominant constituent. The rings seemed smooth, continuous, and unchanging. For more than three centuries, Saturn was the only planet known to possess rings. In fact, when I completed my own Ph.D. research in the mid-1970s, fellow students viewed Saturn's rings as a sort of scientific backwater, lacking for them the captivating excitement of cosmology, galaxies, or exploding stars. One of the major research questions of that time was, "Why is Saturn the only planet with rings?" Several plausible answers were proposed, which now are naturally only of academic interest.

Then came the deluge. Within about a decade, from 1977 to 1989, rings were discovered around all the giant planets and imaged close up by the Voyager spacecraft. I had the good fortune to be involved with the space missions Pioneer Saturn and Voyager, and I chaired the Voyager Rings Science Working Group. Rich Terrile has described the experience as "trying to drink from a fire hose." Others have described it as a golden age of planetary ring exploration. I will indulge the reader by giving a brief summary of the last 40 years of ring studies from my own perspective. This may serve as a partial motivation for the detailed chapters that follow.

2.1 Overview of planetary ring exploration

2.1.1 Discovery of the Uranian rings

In winter 1977, as part of my dissertation, I was observing Saturn's rings by night from Lowell Observatory, using the International Planetary Patrol telescope there. During the days, I wandered into my office and was occasionally distracted by office-mate Larry Wasserman's preparation for a rare event: Uranus was to pass in front of, and thus "occult," a bright star. This stellar *occultation* was designed to measure the exact shape of Uranus and the structure of the top-most layers of the planet's atmosphere. Bob Millis led the team at Lowell (later, the director there) in collaboration with the late Jim Elliot (then of Cornell) and others. After the event, I remember clearly hearing Millis on the local radio, reporting from Australia that Uranus must be surrounded by "thousands of moons," because the star had unexpectedly blinked out numerous times before it was covered by the planet's disk. This showed clearly the presence of solid bodies orbiting Uranus. Within days the full story was clarified by Jim Elliot (see his recollections in Elliot and Kerr (1984)), who was observing the occultation events from the Kuiper Airborne Observatory, a converted airplane equipped with a telescope, flying over the Indian Ocean. Because he was not restricted to telescopes affixed to the ground and was using Wasserman's latest predictions, he was in an ideal position to observe both before and after the planet's occultation of the star. Therefore, he noted that the short interruptions of starlight were symmetric: before and after, thus left and right of the planet (see Figure 2.5). The natural explanation was that the thousands of moons reported by Millis were actually billions arranged in nine rings around the planet.

This discovery was only the first of many surprises over the next five years. Other observers quickly confirmed the findings, and the combination of multiple observations determined the widths and locations of the Uranian rings. Not only were Saturn's rings no longer unique, but the new Uranian rings were surprisingly quite different: they were narrow with sharp edges, and some rings were eccentric. Each of these was contrary to expectations – for reasons that will be discussed in some detail in the chapters that follow. It seems that those scientists studying rings had become satisfied with the simplest possible model. In the coming years, every aspect of that model would be overturned.

Rapidly, Peter Goldreich and his postdoctoral colleague Scott Tremaine (1979) explained the narrow rings with sharp edges based on nearby satellites that hold them in place through gravitational interactions. Although gravity is an attractive force, in this case it can confine the narrow rings, which then spread slowly. The proposed moons were termed *shepherds* because they keep the ring particles together like a flock of sheep. The elliptical rings could be accounted for by including the effect of the gravity of the ring particles themselves. Two themes were obvious in these explanations: that the effect of close-by moons had a strong influence on ring behavior, and that the familiar laws of gravity might provide some unexpected, complicated outcomes.

2.1.2 Pioneer and Voyager discoveries

At this time, the first spacecraft were on their way to Saturn: Pioneer 11 had flown by Jupiter in 1974 and would reach it in 1979; Voyager 1 and Voyager 2 would

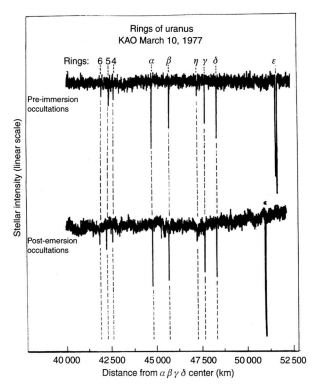

Figure 2.5 Occultations by the rings of Uranus observed with the Kuiper Airborne Observatory on March 10, 1977. Occultations corresponding to the nine confirmed rings are easily seen. Most (if not all) of the low-frequency variations in the lightcurves are due to a variable amount of scattered moonlight on the telescope mirror (from Elliot, 1979). (Reprinted with permission, from the *Annual Review of Astronomy and Astrophysics*, Volume 17, © 1979 by Annual Reviews www.annualreviews.org.)

fly by Saturn in 1979 and 1981 after gravity assists from Jupiter in 1979. At that moment in the late 1970s, only a few workers specialized in the study of planetary rings, and I had the good fortune to be one of them. I was also involved with each of these missions. Like everyone else, I had no expectation of the many surprises headed our way: I, too, was still focused on the simplest ring model.

The next surprises were at Jupiter. Six years earlier, Pioneer had detected a disappearance of *radiation belts* near the planet that could be explained by their being erased at that particular location by absorption due to a Jupiter ring. After some argument, Tobias Owen convinced his colleagues and the Voyager project management to invest precious minutes as the Voyager 1 spacecraft passed over the Jupiter equator to stare at apparently blank space in the direction of a possible ring. This investment paid off when the smeared image clearly showed a fuzzy ring surrounding the planet. When Voyager 2 followed only months later, it was reprogrammed to snap a small number of images, which showed the new ring to be yet more different from the expectation based on Saturn's rings – it was broader than the Uranian ring, but ethereal, a million times less opaque than Saturn's main ring. Furthermore, the Voyager spacecraft were able to image the rings backlit by the Sun after flying by the planet (this perspective is never available from the Earth), and the ring appeared much brighter when viewed at this angle. This phenomenon is due to light diffraction by small particles, showing that the size of most of the particles in Jupiter's ring is near to that of visible light (see Chapter 15). These small ring particles (of size 1 µm or less) are naturally called "dust," although they are closer in size to smoke particles on Earth. Thus, the Jupiter ring showed yet more variety, being the first "dusty" ring. Joseph Burns *et al.* (1980) proposed that the dust arose from small satellites that served as *parent bodies*. These moons would continually resupply the dust, which would be swept away in a cosmic eyeblink of just thousands of years, to fall into Jupiter's atmosphere.

Later that same summer of 1979, Pioneer 11 became the first spacecraft to reach what we had called *the* Ringed Planet just a few years earlier. The late Tom Gehrels, imaging team leader of the Pioneer Saturn mission, had invited me to join his team, with responsibility for ring studies, and assigned me the task of searching for new rings. The close flyby of Saturn allowed us to detect a new ring, which Tom and I decided to name ring F, the next available letter in the alphabet. Rings D and E had already been reported by others, although Pioneer could not see them. We didn't want to confuse the nomenclature, even though Pioneer could not confirm or deny their existence. In the next days, another ring, called ring G, was detected by its absorption of radiation belt particles. Like many others on the spacecraft team, I went without sleep to interpret the new data and present our results to an interested public. Tom Gehrels was very careful to assure that our findings were a real discovery and not some artifact.

Even after these spectacular experiences, we were still unprepared for the variety in the Voyager flyby of Saturn. The Voyager cameras were much superior to those of Pioneer. At some distance from Saturn, a few weeks before closest approach, their capability already exceeded any previously available. On that day, the spacecraft produced the "best ever" picture of Saturn and its rings. Of course, the next day the images were better still. The daily sequence of spectacular new images and discoveries kept Voyager investigators in a continual state of excitement as each new image showed an increase in the number of rings surrounding Saturn. Reporters and the public tried to keep up with the latest number. The highest-resolution studies, which were provided from stellar and radio occultations, eventually showed tens of thousands of features. Naturally, not all of these were designated rings. A list of 216 prominent features is provided by Esposito *et al.* (1987).

The Voyager flybys showed close-ups of the new rings D, E, F, and G; abundant structure in the main rings; waves, *wakes*, and scalloped edges; and numerous small moons. Two of these new moons, now known as Pandora and Prometheus, were on either side of Saturn's F ring, apparently confirming the "shepherding" theory, that the ring is held in place by gravitational interactions with these small moons.

One of the most remarkable discoveries of the Voyager encounters were the numerous waves visible in Saturn's rings. The cameras, the radio science team, the *photopolarimeter* experiment, and the ultraviolet spectrometer all saw ripples passing through the rings, each excited by the gravity of nearby moons. Each wave was generated at a location in the rings where the natural orbital motions of the particles lay in a *resonance* with the motion of a nearby moon. Remarkably, these spiral waves could be explained by the same theory developed earlier to explain the arms of spiral galaxies. These waves would later provide estimates of the rings' total mass, random velocities, thickness, and age.

2.1.3 The theoretical onslaught

The sudden burst of information about Saturn's rings and the variety of new phenomena seen there encouraged an immediate response from physicists and astronomers. Particularly, the theoretical team of Borderies, Goldreich, and Tremaine at the Jet Propulsion Laboratory/California Institute of Technology set out to explain the dynamics. Over the following years, they produced a series of papers that provide our basic understanding of how planetary rings behave. They borrowed freely from the astrophysical literature of galactic dynamics and disks and applied the classical physics of statistical mechanics, orbital dynamics, and fluid dynamics to these current problems. The rapid pace of new discovery and explanation benefited both the theoretical and observational teams. All of us who were involved remember, with pleasure, those exciting days and the continual intellectual interplay.

Not only did the youthful study of planetary rings draw from more classical studies, but the quality of the data and the need for detailed explanations of ring phenomena forced refinement of some astrophysical models. As an example, Frank Shu and his collaborators (Shu *et al.*, 1985a, b) extended the spiral wave theory to the non-linear case and applied the results to galactic dynamics. Bill Ward and his collaborators (e.g., Ward and Hahn, 2000) took a close look at the interactions between disks and embedded moons; this has provided a severe constraint on the evolution of protoplanetary systems around other stars. Like the moons near a planetary ring, the growing *protoplanets* exchange momentum with the disk and evolve by moving radially. In a very short time, these planets can be driven into a fiery death in the forming star.

The quick interaction between observation and theory led to a rapid general understanding of the major phenomena. Much of the excitement and many of the results can be gleaned from the book *Planetary Rings* (Greenberg and Brahic, 1984), resulting from a conference in Toulouse, France, in 1982. Amazingly, much of what we understand today is encapsulated in this seminal University of Arizona Press book. Remarkably few of the ideas are dated. Some workers still refer reverently to this large book as "The Bible." Elliot and Kerr followed up with a popular book in 1984.

Two major developments followed these early golden years of planetary ring studies:

Astrophysics and geophysics. The Voyager flybys of Uranus in 1986 and Neptune in 1989 provided close-up views and a uniform data set on all ring systems. These details have caused a shift from a more "astrophysical" to a more "geophysical" approach. As for many Earth phenomena, the precise details are critically important in understanding the behavior of planetary rings, and the present state is contingent on particular events, sometimes random, in their history. This nitty-gritty work of gathering the detailed data and detailed explanations has proceeded more slowly. One example is the case of missing shepherding satellites. The F ring is not exactly placed where the shepherding *torques* would balance. Of the Uranian rings, shepherds were found only for the largest ε (epsilon) ring; even so, they are too small to hold it in place for the age of the solar system. Another issue is that the sharp edges of rings are too sharp. Progress on each of these problems as well as many others has continued, but more slowly. The detailed models must consider more physics, and in many cases the Voyager data were insufficient to decide the issues conclusively. Cassini is now offering new perspectives and new questions, but these old problems remain unsolved.

Young rings. The second development was the continuing need to explain the rapidity of many ring processes. Diffusion of ring particles, ring spreading, darkening by *meteoroid* bombardment, grinding by collisions and meteoroid impact, *gas drag*, *plasma drag*, and transfer of momentum to shepherding moons all have timescales shorter than the age of the solar system. We were immediately forced to address this particular difficulty (I will discuss this more in later chapters). The many nearby moons discovered by Voyager and others are obvious sources for ring renewal. For the thinner dusty rings of Jupiter, Galileo data confirm their continuing creation. In my 1985 Urey Prize lecture (Esposito, 1986), I proposed that ring systems and particularly Saturn's rings were recently created from the shattering of small moons. The destruction of a moon would simultaneously be the creation of a new ring! This idea is now generally accepted for many narrow rings, but it requires a particularly rare event to account for the recent creation of Saturn's rings. For many, this explanation requires just too special an event: one so rare that we should not be so lucky to observe its short-lived after-effects now. On the other hand, this explanation highlights the importance of random events in ring history. The issue is still not resolved, particularly for Saturn. It seems that we must either find some way to slow ring evolution or postulate that rare and unlikely events account for the rings we now observe. For the latest thinking, see the more detailed discussion in Chapter 10.

2.1.4 Cassini at Saturn: Results and plans

The Cassini orbiter has been orbiting Saturn since 2004, and the mission is currently extended until 2017, when Saturn is at solstice. In 2009, Cassini observed the rings of Saturn at equinox, when they lie parallel to the Sun's rays. At that time, any vertical variations or larger objects catch the light of the Sun and shine like a mountain peak at sunset. The shadows are long and much bigger than the features that cast them: these two effects allowed the Cassini instruments to observe phenomena that were much too small to resolve previously.

Cassini observations confirm that the ring particles are composed of mostly pure water ice, with some contaminants. The particles cover a range of sizes from dust to small moons. Some small embedded moons were discovered by noticing a *propeller*-shaped effect on nearby ring material. Cassini showed that the ring particles form temporary elongated aggregates tens of meters across called *self-gravity wakes*. The rings are highly dynamic, with some changes apparent since the Voyager flybys. A few aspects of ring structure can even change in a matter of days or weeks . . . we literally see the rings change before our eyes. The rapid evolution is hard to reconcile with ancient rings that are as old as the solar system, unless some renewal or recycling is occurring. Cassini will directly measure the rings' mass and the nature of the interplanetary particles that continually bombard them. This will help decide if the rings are remnants of the Saturn nebula or fragments of a destroyed moon or comet.

Cassini has focused its study on many of the dynamic structures that change over timescales of hours, months, and years. Small-scale structure is apparent: gravitational wakes, the surprising phenomenon of *overstability*, and individual objects that are only 100–1000 m in dimension. Vertical structures of kilometer extent are created by small moons on inclined orbits. Satellites open gaps, some of which extend for just a part of the circumference, and produce "propeller" structures. High *phase angle* images show dusty rings that distort under the Sun's influence. Larger objects embedded in the F ring that are only about a kilometer in size stir the ring up, with the objects occasionally colliding to release *jets* of dust.

In these stirred-up regions, the ring particles are agitated, creating higher velocity, but also temporary aggregation. The ring spectra and *photometry* clearly show the effects of this stirring, with bright "haloes" surrounding the strongest *density waves*. There is evidence that the rings of Saturn are older than Voyager suggested, going back to the formation of the solar system or perhaps to the era of late heavy bombardment that followed, about 3.9 billion years ago. Because the rings appear to be nearly pure ice, this might require them also to be much more massive than previously thought, perhaps by a factor of ten. This would explain why they are not more polluted by the continuing infall of meteoritic material on the rings. There is some evidence for more massive rings based on self-gravity wakes seen by star occultations and from re-interpretation of some previous Pioneer 11 results. Further, simulation of particle dynamics by Stuart Robbins in 2010 shows that the particles tend to clump, indicating that they are likely to be much more massive than estimated by Voyager and earlier results (including my own estimates; see Esposito *et al.*, 1983b) that assumed the rings were homogeneous.

I have suggested (Esposito *et al.*, 2012) that the ring dynamics resemble an ecological "predator–prey" system. Cassini has found sporadic F-ring changes, including new transient objects and new ring *strands*; the *moonlets* that created propeller structures change their orbits; interfering wave modes flow along the A and B ring edges; the ethereal *spokes* come and go. Some of this unexpected behavior is driven by deterministic interactions, but other aspects seem to be *stochastic*, the result of random events.

Prometheus' maximum stirring of the F ring was in 2009, which may predict the formation, in the next few years, of larger objects in the ring. Elsewhere in the rings, objects may also be forming where the rings are stirred up by resonances or nearby moons.

When the aging spacecraft later flies very close and even inside the rings, in its final *proximal* orbits, Cassini will systematically survey the Saturn ring system at multiple wavelengths, and provide ultra-high resolution observations by watching occultations of selected stars whose apparent motion nearly tracks the ring-particle orbits. The extended mission will provide a complete seasonal coverage of spokes.

At the very end of the Cassini mission, before it becomes a short-lived meteor in Saturn's atmosphere, many close-up views will be possible: during the last orbits Cassini will speed between the rings and Saturn at its closest approach. The spacecraft will pass so close to the rings that their gravity will change its orbit slightly, allowing us to directly measure, for the first time, the mass of the rings. This key result will test theories of the rings' age and evolution. If we find massive rings, they could be as old as the solar system, but continually changing due to the random events they suffer.

2.2 Summary

From the first, our understanding of planetary rings has been driven by the need to explain our observations. The close interplay of observation and theory continues to the present day. We are in yet another golden age of scientific advance triggered by Cassini observations since 2004. The simple conceptions of 40 years ago have given way to detailed astrophysical and even geophysical models, where random events and the interactions of moons and rings are essential to our understanding. Rings continue as a local laboratory and a paradigm for other flattened systems in our universe. The following chapters provide a current update on our present knowledge of the physics and phenomena of planetary rings.

3

Diversity of planetary rings

Planetary rings are composed of myriad small particles, mostly orbiting inside the *Roche limit* of the giant planets. The Roche limit is the distance closer than which a fluid particle would be disrupted by tidal forces from the planet; at this limiting distance, the tides are just balanced by the self-gravity of the object (Weidenschilling *et al.*, 1984). An interesting exercise is to compare all the ring systems when normalized to the equatorial radius of each planet (see Figure 3.1, after Nicholson and Dones, 1991). This comparison clearly shows that the rings occupy a common location near the planets, overlapping with numerous small moons (called *ring-moons*) near the rings (Thomas, 1989). As we shall see, these moons are dynamically (and likely genetically) related to the nearby rings.

The properties of the individual particles in the rings are poorly known. We have the best information for Saturn's rings, which are the brightest and have been studied from the ground since their discovery by Galileo in 1610 (see, e.g., van Helden, 1984). The most recent data are from the Cassini orbiter mission (Matson *et al.*, 2004; Cuzzi *et al.*, 2010). In the same time period as the Cassini mission, the Spitzer infrared telescope, orbiting the Earth, discovered what may be the largest ring in the solar system, the Phoebe ring. This ring is created by material from the moon Phoebe that gradually spirals inward, perhaps all the way to Saturn's moon Iapetus, where the resulting deposits may explain the origin of Iapetus' dark side (Tamayo *et al.*, 2011). Our current understanding, including the best available inferences, is summarized in Table 3.1 (after Nicholson and Dones, 1991; Burns *et al.*, 2001; Porco and Hamilton, 2007).

A great variety of structure is seen in the various ring systems, arising from internal and external physical processes. Each ring system includes diffuse, tenuous rings, which Burns *et al.* (1984) have termed "ethereal" rings. Jupiter (Figure 3.2) provides the best studied example. The Jupiter rings are composed of small dust grains released by the intermingled moons Adrastea, Metis, Amalthea, and Thebe. The densest ring of Jupiter, called the "main" ring, extends inward from the

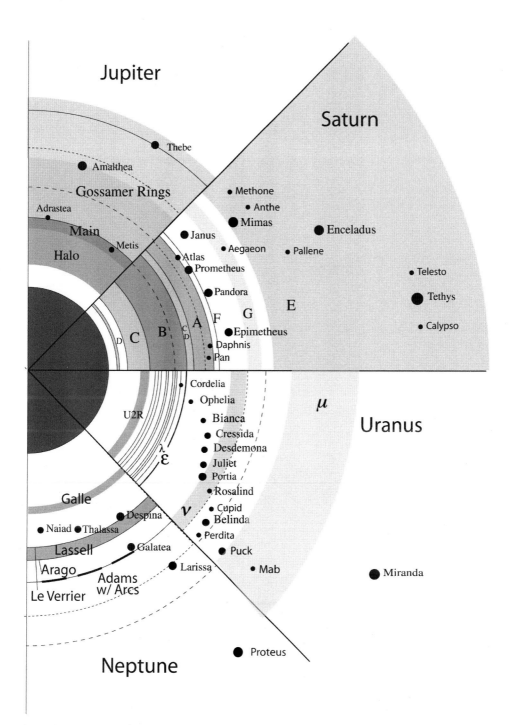

Figure 3.1 A comparison of the four planetary ring systems, including the nearby satellites, scaled to a common planetary equatorial radius. Density of shading indicates the relative *optical depth* of the different ring components. Synchronous orbit is indicated by a dashed line, the Roche limit for a density of $1\,\mathrm{g\,cm^{-3}}$ by a dot–dash line. (Figure courtesy of Judith K. Burns, updated by Doug Hamilton and by the author. From Burns *et al.* (2001), Dusty rings and circumplanetary dust. In *Interplanetary Dust*, ed. E. Grün, B. A. S. Gustafson, S. F. Dermott, and H. Fechtig. Berlin: Springer-Verlag, pp. 641–725; reproduced by permission of the publisher.)

Table 3.1 *Planetary rings characteristics*

	Location (width)	Optical depth	Dust fraction (%)	Power-law index	Notes and vertical thickness
Jupiter					
Halo	92 000–122 500 km	10^{-6}	100	?	12 500 km thick
Main ring	122 000–128 980	3×10^{-6}	~50 (?)	$q \leq 2.5$	Bounded by Adrastea
Amalthea Gossamer	129 000–182 000	10^{-7}	100 (?)	?	2000 km thick
Thebe Gossamer	129 000–226 000	3×10^{-8}	100 (?)	?	4400 km thick
Saturn					
D ring	66 000–74 000	10^{-3}	5–100	?	Internal structure
C ring	74 490–91 983	0.05–0.5	<3	3.1	Some isolated ringlets
B ring	91 983–117 516	≤2.5	<3	2.75	Abundant structure
Cassini Division	117 516–122 053	0.05–0.15	<3		Several plateaus
A ring	122 053–136 774	0.65	<3	2.75–2.90	Many density waves
F ring	140 200 ($W \cong 50$ km)	0.1–0.5	>98	2–3	Narrow, broad components
G ring	166 000–175 000	10^{-6}	>99	1.5–3.5	Arc, coincident with Aegaeon
E ring	180 000–1 200 000	10^{-5}	100		Peak near Enceladus
Phoebe ring (?)	60–400 R_s ($R_s = 60 330$ km)	2×10^{-8}	100	?	The particles may coat Iapetus leading hemisphere
Uranus					
1986 U2R	37 000–39 500	10^{-4}–10^{-3}	?		Still unnamed but may be the same as ζ
ζ	37 850–41 350	10^{-3}		?	

Ring	Distance (km)			q	Notes
Dust belts	41 000–50 000	1–10^{-5}	?	?	Fine internal structure
6	41 837	0.3	<1	$q > 3.5$	
5	42 234	0.5	<1	$q > 3.5$	
4	42 570	0.3	<1	$q > 3.5$	
α	44 718	0.3	<1	$q > 3.5$	
β	45 661	0.2	<1	$q > 3.5$	
η	47 175	0.3	<1	?	
γ	47 627	2	<1	?	
δ	48 300	0.4	<1	?	
λ	50 023	10^{-3}	>95	?	Faint, dusty ring
ε	51 149	0.5–2.3	<1	$2.5 < q < 3.0$	Adjacent to Cordelia
ν	66 100–69 900	6×10^{-6}	<1	?	Between Portia and Rosalind
μ	86 000–103 000	8×10^{-6}	<1	?	Coincident with Mab
Neptune					
Galle	41 000–43 000	4–10×10^{-5}	?	?	
Le Verrier	53 000 ($W = 10$ km)	10^{-2}	4–70	?	Adjacent to Despina
Lassell	53 000–58 000	1–3×10^{-4}	?	?	
Arago	57 200 ($W = 100$ km)	?	?	?	
Adams	62 930 ($W = 50$ km)	10^{-2}	2–50	?	Adjacent to Galatea
Adams arcs	62 930 ($W = 10$ km)	10^{-1}	4–70	?	

Sources: After Burns *et al.*, 2001; Nicholson and Dones, 1991; French *et al.*, 1991; Porco and Hamilton, 2007.

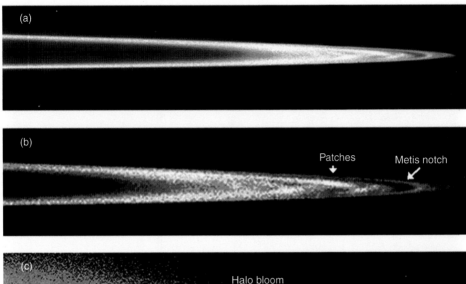

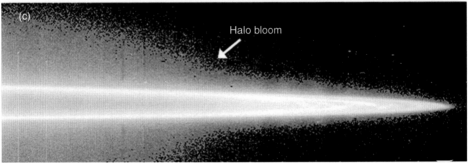

Figure 3.2 A Galileo view of the Jovian ring, showing both the main ring and the halo's outer parts, processed in three different ways to highlight various features. (a) Stretched to differentiate the main ring's diffuse inner periphery versus its much crisper outer boundary. (b) A stretch that emphasizes the patchy nature of the main ring's central region located just interior to a brightness dip associated with Metis' orbit. Features that are bright just above a horizontal line through the ring's edge tend to become dark just below the line, and vice versa. (c) By emphasizing fainter structures, the halo's development at the main ring's inner edge is revealed; it appears that the main ring itself is enshrouded in a faint cloud of material, the so-called "halo bloom," above and below. (From Ockert-Bell *et al.* (1999); reproduced with permission from Elsevier.)

approximate location of Adrastea. Metis orbits closer, within the ring, and may be responsible for a lane of decreased brightness called the *"Metis Notch."* Much of the ring is composed of small, short-lived dust particles with size $1 \leq r \leq 15\,\mu\mathrm{m}$. Above and surrounding the Jupiter main ring is a *toroidal* distribution of dust called the "halo" (Figure 3.3). Outside the main rings, we find two very faint *gossamer* rings (Figure 3.4), associated with the moons Amalthea and Thebe: Galileo observations confirm that each of these small moons is the source of the

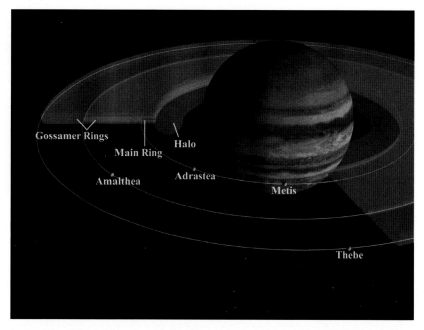

Figure 3.3 A cut-away view of the components of Jupiter's ring system is shown in relation to Jupiter and its small ring-moons. The innermost and thickest ring, shown as a torus, is the halo whose outer edge ends at the narrow and flat main ring. The main ring is circumscribed by the satellite Adrastea's orbit; it may be partly composed of fine particles knocked off Adrastea, and a somewhat larger moon Metis located about 1000 km closer to the planet. Thebe and Amalthea, satellites that are larger still, supply dust that forms the thicker, washer-like gossamer rings; the thicknesses of the gossamer rings are determined by the inclinations of these two satellite orbits. A very faint extension (shown in Figure 3.4) of the outer gossamer ring reaches beyond Thebe's orbit. (From Ockert-Bell *et al.* (1999); courtesy NASA/JPL/Cornell University.) For color version, see Plates section.

ring material interior to it (see Chapter 14). The Thebe gossamer ring extends slightly outward from Thebe's orbit, perhaps due to charging and discharging as ring particles go in and out of Jupiter's shadow. The *Jovian* ring particles probably have silicate compositions, like the surfaces of the nearby moons. Even the biggest objects in Jupiter's ring system are likely to be less than 1 km across.

The rings of Uranus are narrow, with many eccentric, and some also inclined. They include dense rings with sharp edges unlike Jupiter's diffuse rings. Their composition is not known, but their surfaces can not be dominantly water ice (like Saturn's rings) because their reflectivity is so low.

Neptune's rings resemble the Uranus rings: they are primarily narrow, although further from the planet and also dustier and not so opaque. The rings are named

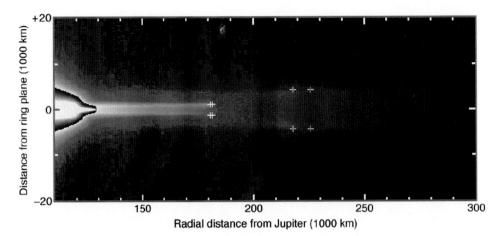

Figure 3.4 This mosaic of four Galileo images (416088922–416089045), taken through the clear filter (0.611 μm) at an elevation of 0.15°, shows the edge-on gossamer rings of Jupiter across phase angles of 177–179°. The halo and main ring are overexposed (solid white with a black outline; cf. Figure 3.3) at left. White crosses mark the extremes of the radial and vertical motions of Amalthea and Thebe as caused by their eccentric and inclined orbits. Amalthea (whose position is roughly in mid-image) bounds one gossamer ring (its ring is the narrower and brighter strip extending to the right from the main ring); Thebe's ring is the thicker and fainter band reaching yet further right. A very faint outward extension to the Thebe ring is also apparent. This image has been enhanced logarithmically to show all the ring components; in reality the Amalthea ring is approximately 10 times fainter than the main ring, while the Thebe ring is 10 times fainter again than the Amalthea ring. Note that each gossamer ring is densest along its vertical extremes, particularly the top strip of Amalthea's ring. The image has been expanded vertically by a factor of 2 to better show the rings' vertical structure. (From Burns *et al.* (2001) Dusty rings and circumplanetary dust. In *Interplanetary Dust*, ed. E. Grün, B. A. S. Gustafson, S. F. Dermott, and H. Fechtig. Berlin: Springer-Verlag, pp. 641–725; reproduced by permission of the publisher.)

for individuals associated with the discovery of the planet Neptune. They exist intermixed with numerous small moons, further from the planet than the other known ring systems. Neptune was the first to show longitudinally incomplete rings, called ring *arcs*.

Uranus has 10 narrow, sharp-edged rings, with eccentric shapes and small inclinations (Figure 3.5). Both the narrowness and sharp edges can be explained by the confining action of nearby moons, called shepherds. For the outermost of these narrow rings, the ε ring, these moons were photographed by Voyager in 1986 (Figure 3.6). For the other rings, the shepherds are still only hypothesized, not yet discovered. The eccentric shape may be explained by the rings' own gravity, or perhaps by particular characteristics of the ring particles' collisions at preferential

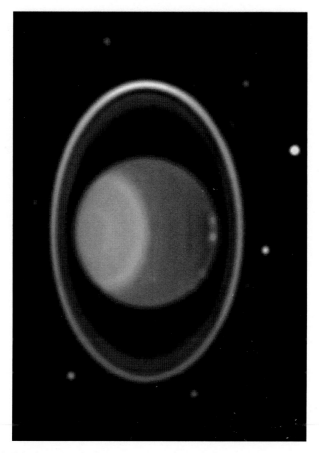

Figure 3.5 This image of Uranus shows the planet and some of its moons in infrared light. Its rings are clearly visible, and their orientation is unusual. Unlike most planets, whose poles point perpendicular to the Sun, Uranus and its rings rotate around an axis that is nearly in the plane of its orbit. (Courtesy NASA/JPL-Caltech.)

longitudes. Lifetimes of the Uranian rings are short, due to the inward drag from Uranus' extended atmosphere, and due to erosion by charged particles for the outer rings.

The spaces between the Uranian rings are filled by broad, diffuse rings of small particles. These were only visible when Voyager 2 turned its camera back toward the planet after passing Uranus and observed the backlit Uranian ring system. This phenomenon is characteristic of particles whose size is approximately the same as the wavelength of light they are seen with (see Chapter 15). The extensive sheets of material are likely to be debris knocked off from (yet undiscovered) small moons among the rings.

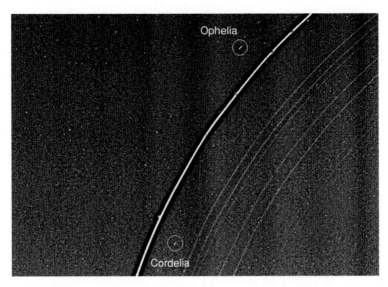

Figure 3.6 Cordelia and Ophelia, a pair of shepherding satellites on each side of the Uranus ε ring, are thought to keep the ring particles in place through resonant gravitational forces. No other shepherds for the other rings were detected by Voyager 2 during its 1986 flyby. (From PIA01976; courtesy NASA/JPL-Caltech.)

Neptune's ring system includes both broad and narrow rings (Figure 3.7). The rings are interspersed with four small satellites with radii of 30–80 km. The gravitational effects of the moon Galatea are clearly evident on the Adams ring (Neptune's largest), and may possibly maintain its hyphenated longitudinal structure (see Chapter 13). The first stellar occultation searches (which observed stars as they passed behind Neptune) were carried out from the ground. They gave inconsistent results: sometimes the starlight was blocked by a ring, and sometimes not. The proposed solution, that Neptune's rings are discontinuous, consisting of a series of incomplete "arcs," was spectacularly confirmed by Voyager 2 in 1989. These arcs are embedded in a diffuse complete ring almost invisible from Earth. The three main arcs are known as Liberté, Egalité, and Fraternité from the call to arms of the French Revolution. A fourth, later discovered, arc is known as Courage, pronounced with a French accent on the second syllable. The brightest Neptune rings are named after the predictors and discoverer of the planet Neptune: Adams, Le Verrier, and Galle, respectively. Lassell and Arago are, respectively, British and French astronomers of the nineteenth century. William Lassell discovered Neptune's moon Triton. Arago measured the diameters of the planets, and as director of the Paris Observatory, was embroiled in the controversy over credit for Neptune's discovery (Moore, 1995).

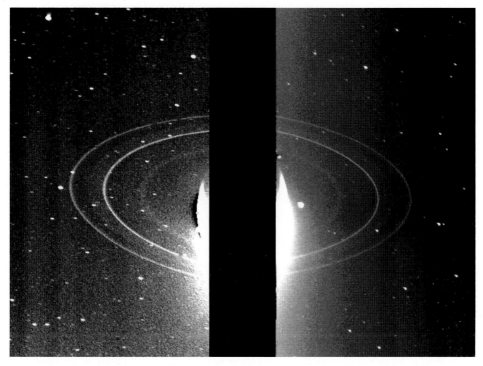

Figure 3.7 A pair of Voyager 591-second exposures (FDS 11446.21 and 11448.10) taken 1.5 hours apart through the clear filter of the Voyager wide-angle camera at a phase angle of $\varphi \sim 134°$, that is, looking back toward the Sun (Smith *et al.*, 1989). The arc region of the outermost Adams ring was not captured in either image. (PIA01997; courtesy NASA/JPL-Caltech.)

Saturn's rings are the biggest and brightest in the solar system (Figure 3.8). They contain as much mass as the moon Mimas, and display all the phenomena found in the other three smaller ring systems. This includes gaps with embedded moons and *ringlets*, narrow rings, broad rings, ethereal rings, waves, wakes, and wiggles (see Chapter 7). Voyager saw time-variable radial dark lanes on the B (brightest, broadest) ring which were called spokes; density waves in the outer A ring; it saw the planet through the partly transparent C (crepe ring); and took close images of D, E, F, and G rings. The alphabetical naming of Saturn's rings, which follows the chronological discoveries of new rings, failed to keep up with the immense number of features seen by Voyager. Now, most ring features beyond the classical, pre-Voyager, rings are identified merely by their location. Ring D lies inside the brighter rings; ring E is a broad, tenuous ring centered on the moon Enceladus. The F ring (shown in Figure 3.8) is a narrow ring just outside the A ring, discovered by the author during the Pioneer 11 flyby in 1979. The G ring

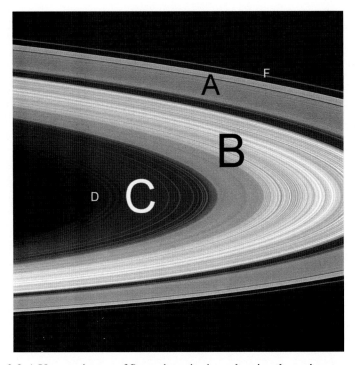

Figure 3.8 A Voyager image of Saturn's main rings showing the main components: A, B, and C. The C ring is the innermost and most optically thin component. The B ring is the middle and densest component. The A ring is the outermost component. It is separated from the B ring by the Cassini Division. The very narrow F ring is seen beyond ring A. The globe of Saturn is extremely overexposed to bring out the details in the rings, a circumstance that enhances the shadow cast by the rings across the equatorial region of the planet. (Courtesy NASA/JPL-Caltech.)

(Figures 3.9, 3.10) is another ethereal ring outside ring F. Cassini entered the Saturn system in an apparently empty area between the F and G rings in July 2004. A mnemonic for remembering the order of the rings from the outermost to the innermost was devised by my students in Introductory Astronomy at the University of Colorado: "Esposito Great For Astronomy, But Can't Draw," apparently a comment on my blackboard skills.

Many small moons are also found among the rings. Figure 3.11 shows a close-up of the small moon Epimetheus. The effects of these moons in Saturn's rings are obvious: they shepherd the F ring, scallop the edge of the A ring, excite hundreds of density waves in the rings, maintain the Encke Gap and the Keeler Gap in Saturn's A ring (see also Figure 7.7). In later chapters I discuss these physical interactions in greater detail. Ultimately, the ring-moons may also provide the material for future rings and be the repository of past ring material (see Chapter 10).

Figure 3.9 Saturn's G ring as seen in a high-resolution Hubble image taken on October 14, 1996. The top right region displays the background-subtracted G ring itself. For guidance, the bold white lines mark (in order of increasing radius) the A ring outer edge, the F ring, and the mean radius of the G ring. The two lines surrounding the G ring represent circles in Saturn's ring plane with radii of $r = 165\,000$ and $175\,000$ km. The trailed image of the moon Epimetheus is encircled and scaled in intensity for visibility. The visibility of the G ring is improved, at the expense of resolution, by smoothing the data with a 3×3 boxcar filter, as shown in the upper left. (From Lissauer and French (2000); reproduced with permission from Elsevier.)

Now that hundreds of planets have been found around other stars, we can ask if any of these exoplanets have moons or rings. To date, neither has been confirmed. The technique of detecting planets as they transit in front of a star, which is the way most exoplanets have been found, could also provide signatures of their rings. If the exoplanets found so far resemble the outer planets of our solar system, they

Figure 3.10 Diffuse rings of Saturn. This image was taken with the Sun almost directly behind Saturn, showing the G ring, Pallene ring, E ring and Janus/Epimetheus ring. (Image number PIA08328 from the Cassini wide-angle camera on 15 September 2006; courtesy NASA/JPL/Space Science Institute)

would be likely to possess ring and moon systems. Unfortunately, most other planet systems do not resemble ours; they have giant planets that are found orbiting close to their stars. Because these "hot Jupiters" are so close to their parent stars, rings would be very hard to detect (see the analysis by Schlichting and Chang, 2011). Many of the ring systems would be viewed edge-on from Earth; particles in the rings would be quickly dragged away; and the regions close to the star are hot enough to vaporize ice and even melt rock.

Astronomers may have found one planet with a possible ring system: Fomalhaut b, which is 115 times more distant from its star than the Earth is from the Sun. Fomalhaut is a star about 25 light years distant. This planet was the first one ever directly seen around another star. After its discovery, the existence of this planet itself was even in doubt, because it was not found by Spitzer follow-up observations in the infrared. Fortunately, recent Hubble observations show that the planet does exist, earning it the appellation "Zombie Planet" after its reappearance.

If the planet has rings, this may explain why it is so bright in visible light, but dim in the infrared: the rings would reflect sunlight, making the planet appear brighter.

Figure 3.11 Close-up of Saturn's small moon Epimetheus, showing its rugged surface, characteristic of the small irregular moons of Saturn that orbit near and within the rings. (PIA09813; courtesy NASA/JPL-Caltech/Space Science Institute.)

To be so bright, much of the ring would need to be outside the Roche limit, and it should quickly collapse to form moons.

A second possible planetary ring system was found by the SuperWASP (Wide Angle Search for Planets) telescopes in the United Kingdom. Even though the planet that the rings surround was not seen, the observed dimming of the star could have been caused by an immense ring system that blocked the star's light for almost two months. It would have gaps in it, to explain the starlight variations, and would stretch tens of millions of kilometers in diameter. Since the planet the rings may surround is undetected, that object might actually be a brown dwarf star and not a planet. The gaps could be caused by embedded objects or resonances with external moons (see Chapter 6). If the unseen planet is a "Super-Jupiter" or even bigger, its moons could be as large as a small planet themselves.

Summary

Planetary ring systems in our solar system show surprising variety. All are formed of immense numbers of small particles, individually too small to see, but collectively appearing as a huge array of structures circling the giant planets. Particles range in size from dust to small moons. The composition of the ring particles, where we can determine it, resembles that of the nearby moons. Some rings are narrow, others broad; some eccentric, inclined, and partial; gaps in rings are empty, or occupied by small moons. The largest ring in the solar system was recently discovered by the Spitzer Space Telescope and named the Phoebe ring. Its particles probably orbit Saturn in a *retrograde* direction, as does Phoebe itself. The inwardly spiraling particles eventually hit Iapetus, darkening its leading side. Moons sculpt and confine rings, providing the sources and perhaps sinks for ring material. Saturn's rings encompass the phenomena seen in other rings and more. In the following chapters, I will explain the plausible causes of ring structure and the source of the incredible variation we observe in planetary rings.

4

Individual ring particles and their collisions

From Chapter 3, we recall that planetary rings are composed of myriad individual particles. Because each particle is on its own individual orbit, potentially intersecting others, it is natural that the particles interact with each other, and that they continually collide. Because no one has ever seen an individual ring particle, we must indirectly determine the characteristics of these individual satellites, and infer the outcome of the collisions. A growing consensus is that ring particles are actually agglomerates of smaller elements that are at least temporarily cohering: these temporary bodies are subject to both growth and fragmentation. The balance between the competing processes yields both a distribution in size and a distribution of velocity. In this chapter we look in more detail at the particle collisions and balance between *accretional* growth and disruption.

4.1 Collisions

When two ring particles collide, the outcome depends not just on their relative motions, but also on their physical characteristics. The exchange of energy and momentum in the impact yields new directions and velocities (or, equivalently, new orbital elements) for the particles afterwards. The energy balance due to collisions determines the overall structure of the ring system, and thus its overall thickness, which is like a temperature. Planetary rings are very thin, and thus "cold." The most basic issue that concerns us about the collisions between ring particles is the amount of kinetic energy dissipated. This loss of energy during a collision is described by the coefficient of restitution, which is defined such that, in an *inelastic* collision, the normal component of the relative velocity is reduced by a factor ε,

$$\vec{v}'_{\text{rel}} \cdot \vec{k} = -\varepsilon(\vec{v}_{\text{rel}} \cdot \vec{k}) \tag{4.1}$$

The left side is the normal velocity (that is, the velocity perpendicular to the particle surface at the point of contact) after the collision, where $\vec{k}$ is the vector connecting the particle centers. On the right side, $\vec{v}_{\text{rel}}$ is the relative velocity before the collision. The partly elastic (that is, $\varepsilon < 1$) collisions remove kinetic energy from the system, which is transformed into heat or deformation of the ring particles. If rebound velocities are sufficiently low, *accretion* can occur. Once smaller particles have accreted onto a larger one, they create a surface layer called a *regolith*; for a particle covered by a regolith, the coefficient of restitution is much lower than for a solid piece of the same material. Thus, the collision characteristics may evolve with time. A similar result can occur as frost forms on particle surfaces or is removed or compressed by collisions as shown in the laboratory experiments of Frank Bridges *et al.* (1984).

Collisions directly affect the size distribution of the ring particles: those collisions that result in sticking (or, equivalently, accretion) reduce the number of small particles and increase the number of larger particles; disruption has the opposite effect. It goes without saying that the likelihood of these different outcomes is strongly related to the collision velocity. It is easy to see that this velocity is determined, in turn, by the amount of dissipation in the collisions, which provides an indirect feedback. In equilibrium, these two processes just balance: the collisional velocity is just large enough that dissipation is *lossy* enough to keep the mean velocity constant. If the ring system is subject to external *forcing*, the system may not always be at equilibrium, but cycle around it.

Two approaches have been used to estimate the ring particles' coefficient of restitution. The first is direct laboratory measurement of collisions between icy bodies at the low velocity characteristic of the gentle collisions in real rings (typical speeds are mm s^{-1}). These even include some space experiments in *microgravity* emulating real ring particle surfaces (Colwell and Taylor, 1999; Colwell, 2003; see Figure 4.1) and some involving large or double pendulums to achieve the gentle impacts (Bridges *et al.*, 1984; Dilley, 1993; Dilley and Crawford, 1996). The second approach involves calculation of the mechanical properties of the small particles (e.g., Hertzsch *et al.*, 1995). Both show a generally declining coefficient of restitution with increasing velocity, which is the basic requirement for thermal stability of the rings (see Stewart *et al.*, 1984; see Figures 4.2 and 4.3). For stability, the thickness or temperature of the rings must not continually increase with collisions; such behavior would cause the ring to "evaporate." At very low velocities, we find that the collisions are nearly perfectly elastic ($\varepsilon = 1$). One complication with this natural expectation, though, is the possibility of thin, cushioning surface layers or even sticky surfaces (Spahn *et al.*, 2004). Thus, collisions at very low (but not necessarily improbably low) speeds can be very lossy, allowing aggregation, or possibly bi-modal collision outcomes where the particles may either stick or rebound. The full implications of these possibilities are unclear for the dense parts

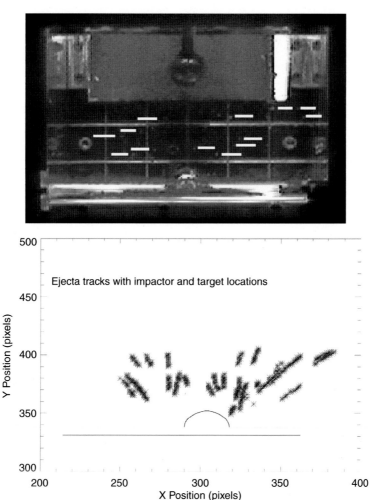

Figure 4.1 A superposition of the trajectories of individual particles and clumps of particles ejected in an experimental impact simulating a collision in a planetary ring aboard the Space Shuttle (top), with a target surface and post-impact projectile position indicated in the lower figure. Horizontal bars in the top figure indicate the individual ejecta. In the lower figure, positions from a series of exposures are overplotted. (From Colwell (2003); reproduced with permission from Elsevier.)

of planetary rings, where collisions are common and velocities are likely to be small. Nonetheless, it is clear that the nature of the particle collisions allows the rings to behave as a very unusual fluid. Although we can derive bulk properties characteristic of this fluid, the derived *transport coefficients*, such as *viscosity*, are not applicable to all situations. Frank Spahn (Spahn *et al.*, 2001) states that "viscosity cannot be considered as a material-specific value: it is sensitively dependent

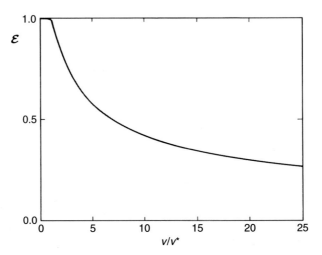

Figure 4.2 The coefficient of restitution as a function of impact velocity. (From "Unresolved problems in planetary ring dynamics" by Nicole Borderies, Peter Goldreich, and Scott Tremaine, in *Planetary Rings*, edited by Richard Greenberg and André Brahic, © 1984 The Arizona Board of Regents. Reprinted by permission of the University of Arizona Press.)

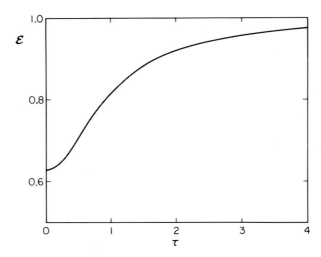

Figure 4.3 The equilibrium value of the coefficient of restitution as a function of optical depth. (From "Unresolved Problems in Planetary Ring Dynamics" by Nicole Borderies, Peter Goldreich, and Scott Tremaine, in *Planetary Rings*, edited by Richard Greenberg and André Brahic, © 1984 The Arizona Board of Regents. Reprinted by permission of the University of Arizona Press.)

on the physical environment." Numerical simulations by Lewis and Stewart (2000) and by Robbins *et al.* (2010) for *perturbed* rings show clear deficiencies of the hydrodynamic description and unexpected structure and phenomena (see Figures 4.4 and 4.5).

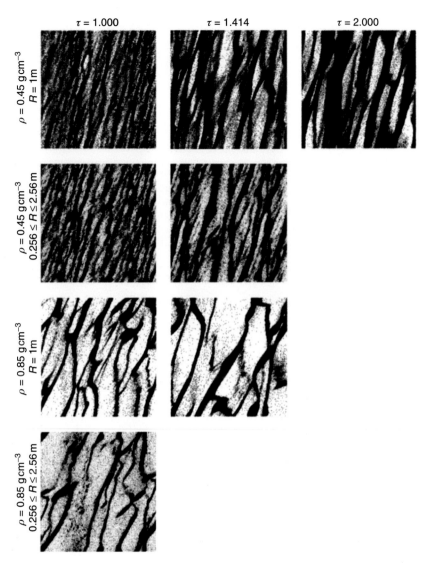

Figure 4.4 Simulation of a real ring does not show uniform structure. Instead a web-like pattern is seen, with some empty regions, and others very dense. The figure shows a variety of B ring simulations for different values of optical depth, particle density and size distribution. (From Robbins *et al.* (2010); reprinted with permission from Elsevier.)

4.2 Ring particle size and shapes

Although J. D. Cassini suggested that rings are actually a swarm of small satellites, and Maxwell proved this with his prize essay in 1857, the individual nature of these small particles was not known. A simplifying tendency is to imagine the rings are

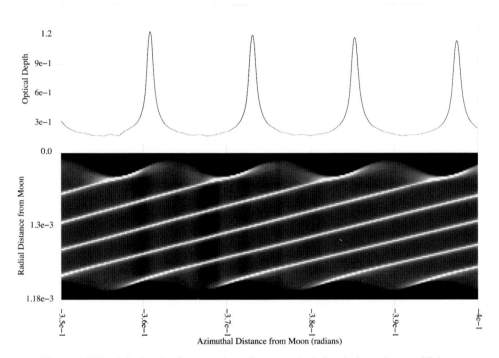

Figure 4.5 Particle density for a section of a numerical simulation where collisions were considered. The top plot shows the data from a cross-section of those data. (The bottom plot is a surface plot, which roughly approximates what would be seen in observations.) These results show sharp peaks in the concentration of ring particles downstream from a nearby perturbing moon. (From Mark Lewis (2001), reprinted with permission from Mark Lewis.)

made of spherical particles all of the same size. Observations of the rings using a given wavelength of light or other part of the electromagnetic spectrum tend to detect particles whose size is larger than that wavelength of observation. Thus, each observer might find a different part or aspect of the size distribution, like the story of the blind men studying the elephant. We now know that the ideas of a single size and shape are completely outdated by the measurements of the last few decades. The so-called "typical size" has tended to grow as longer wavelengths are used to probe the rings. Our current understanding is that the size of ring particles extends over many decades, from fine dust to embedded moonlets, kilometers across. The most natural way to characterize these broad distributions is with a *power law*, that is, the number of particles per unit volume in a small size interval da, centered on size a:

$$N(a)da = C_0 a^{-q} da \quad \text{for } a_{\min} < a < a_{\max} \tag{4.2}$$

where C_0 is a constant related to the total opacity, and $a_{\min}$ and $a_{\max}$ are the smallest and largest particles in the distribution. The *power-law index* is q. On a log–log plot, this distribution would be a straight line with slope $-q$. Typical values of q are around 3, which is also characteristic of the asteroid belt and of size distributions created by shattering objects in the laboratory. These similarities are not likely to be coincidental: both the asteroids and particles in planetary rings were probably created by fragmentation of larger objects and were subject to subsequent collisional evolution. Some groups of objects in the ring show much steeper size distributions, with larger values indicating different histories (see Sremcevic *et al.*, 2007).

Numerical simulations show that the collisions between particles tend to equalize or share the energy between the impactors. This tendency is called *equipartition* of energy. In ideal gases, this is achieved at the state of thermal equilibrium. For rings, this state is only partly reached: the smaller bodies have only 2% to 20% of the kinetic energy of motion of the largest (Salo *et al.*, 2001). However, because of their much smaller masses, the small ring particles have significantly higher velocities relative to a purely circular orbit. These larger velocities represent larger eccentricities and inclinations and are equivalent to a higher temperature that causes their vertical excursions also to be larger. Thus, the particle size distribution leads to a vertical gradient in particle size: the largest particles, with most of the mass of the ring system, are confined to a narrow layer (or a single layer, termed a *monolayer*) in the ring plane, while the smaller particles extend to higher altitudes. The finest particles dominate the distribution above and below the ring plane.

4.3 Dynamic ephemeral bodies or "rubble piles"

The collisions of the ring particles can cause them either to grow in size or to be disrupted. The dynamic balance between these competing processes establishes an equilibrium state of aggregate bodies that resemble piles of rubble. Particles tend to gather together, quickly growing to sizes that resist tidal disruption, only to be broken apart by mutual collisions. Since relative velocities are low and collisions are inelastic, accretion is very rapid. Large particles can hold smaller ones on their surfaces by their mutual gravitational attraction (Canup and Esposito, 1995) or by adhesion (Albers and Spahn, 2006). In Saturn's rings, the timescale may be less than a day for house-sized objects to accrete. After rapid growth beyond several meters, ring particles become increasingly prone to disruption (Weidenschilling *et al.*, 1984). These large *rubble piles* are indeed dynamic and ephemeral: such rubble piles are a complete contrast to the simple idea of a spherical ring particle of a uniform size. An artistic conception of the particles in Saturn's rings is shown in

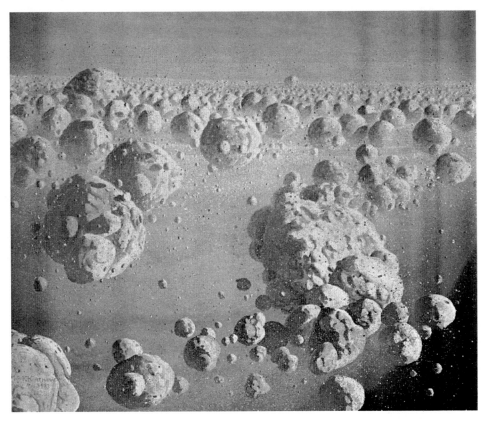

Figure 4.6 *Dynamic ephemeral bodies* (DEBs) in Saturn's rings viewed from just above the plane. House-sized bodies (the larger bodies in the picture) grow in a matter of days by accretion of much smaller particles, shown as a haze among the larger bodies. The large bodies continually break up due to tidal forces, as in the case of the S-shaped disaggregating swarm in the right foreground. Large bodies are irregularly shaped and lie roughly in a monolayer; smaller particles lie in a many-particle-thick layer. This painting represents a thinly populated locale, where optical thickness $\tau < 1$. Zones with $\tau > 1$ would look similar except that the haze of small particles would hide more completely the background. (Painting by W. K. Hartmann, based on the model of particle properties and dynamics discussed by Weidenschilling *et al.* (1984).)

Figure 4.6. A computer simulation is shown in Figure 4.7, where particles cluster around an embedded moonlet.

4.4 The Roche limit

Edouard Roche (1849; see Chandrasekhar, 1969) calculated the distance at which a purely fluid satellite would be pulled apart by tidal forces. This is the distance

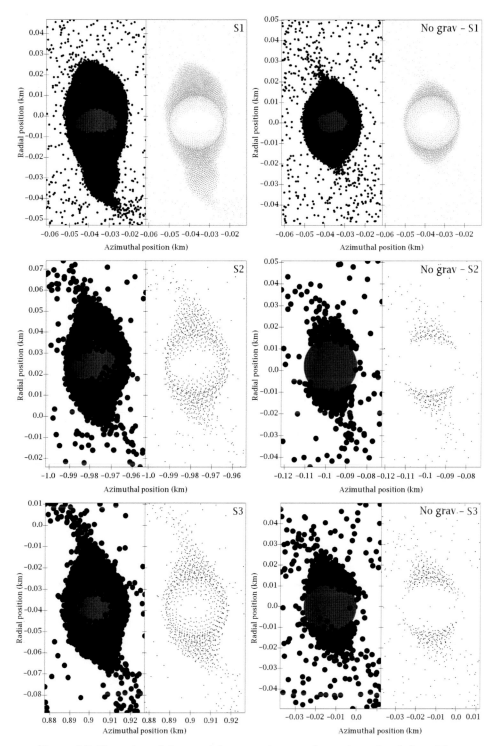

Figure 4.7 Clustering of ring particles around a moonlet, various simulations. The ring particles are shown as small dots on the left panel, and drawn to scale in the right panel for each simulation. On the right, the moonlet is drawn in a lighter shade, to clearly show it is covered by smaller ring particles. (From Lewis and Stewart (2009); reprinted with permission from Elsevier.)

where the gradient in a planet's gravitational force overcomes the gravitational attraction of the satellite's own material: its self-gravity alone is insufficient to hold it together. Of course, solid objects (and we humans, for example) can exist inside the Roche limit without being disrupted by the Earth's tides due to the material strength of their bodies. Even loose aggregates would possess some strength.

Roche's criterion can be written as:

$$\frac{\alpha_r}{R} = 2.456 \left(\frac{\rho_p}{\rho} \right)^{1/3} \tag{4.3}$$

where fluid objects would suffer tidal disruption inside the Roche limit, α_r, for a central planet with radius R and density ρ_p. The particle's density is ρ. Thus, more dense objects could avoid tidal disruption closer to the planet (smaller α_r). Further, for real bodies, stripping of loose material or fracture by tidal stresses occur much closer to the planet than in equation 4.3. This equation can also be inverted to give a *critical density* at each distance: objects with smaller density are disrupted, while those with greater density survive.

Robin Canup and Esposito (1995) numerically characterized how accretion can occur inside the classical Roche limit if one body is much larger than the other. In fact, if they are not rotating, a small body on the surface of a larger one is held by its gravitational attraction if

$$\frac{\alpha}{R} \geq 1.26 \left(\frac{\rho_p}{\rho} \right)^{1/3} \tag{4.4}$$

This region surrounding the classical Roche limit, where the mutual gravitational attraction of two bodies is comparable to their mutual gravitation, is called the *Roche zone*. This is the same region where accretionary growth must compete with tidal disruption, so formation of natural satellites around the planet would also be impeded there.

4.5 Ring particle composition and temperature

Without directly sampling any ring particles, reflection spectra and color give some indication of their composition. In general, the ring particles are similar to the nearby moons. Saturn's rings are predominantly water ice, Uranus' are dark, Jupiter's are derived from nearby Thebe and Amalthea. Color variations across Saturn's rings may indicate varying composition, perhaps because of the interplanetary dust that bombards them and darkens the particles. Saturn's ring particles have rough, irregular surfaces resembling frost more than solid ice. There is good indication that the particles are under-dense ($\rho \leq 1$), further supporting the idea of ring particles as temporary rubble piles. These slowly spinning particles collide gently with collision velocities of just millimeters per second, except possibly where they are stirred up by moon resonances or larger embedded bodies.

Even before Cassini, Saturn's rings were known to be predominantly made of water ice (Esposito *et al.*, 1984), with the water fraction estimated to be 90–99%. However, this ice is impure and darkened by other constituents. Cuzzi and Estrada (1998) explain the colors of Saturn's rings seen by Voyager as due to initially quite pure water ice contaminated by a small amount of organic contaminants, which has been subjected to bombardment by meteorites and micrometeorites over the lifetime of the rings. This dark, relatively colorless material collects in the rings, partly covering the surfaces of the ring particles and occasionally is splashed off by later meteorite impacts. This can explain the relative color of the various rings of Saturn: the optically thinner ring C and Cassini Division have less mass per area and therefore become more polluted by the in-falling meteoritic material. The optically thicker rings A and B have higher mass surface density and would darken more slowly: they remain the brighter ring elements. This model also explains the color variation across the ring boundaries, where splattering by further impacts redistributes the contaminants.

Because of the close range of the Cassini spacecraft just after the insertion into Saturn orbit, the ultraviolet and infrared spectrometers measured ring-reflected sunlight from the rings with a spatial resolution 100 times better than available previously from Voyager, ground-based, and Hubble observations. The reflected sunlight contains the spectrum of absorbing constituents on the ring particles, and consequently these observations measure their ring composition and its variation across the rings. Both ultraviolet and infrared spectra show the absorption features of the water ice, confirming this as a major constituent of the rings (Esposito *et al.*, 2005; Brown *et al.*, 2006). For example, the ultraviolet reflectance spectrum for the rings displays a broad absorption feature near wavelength 1600 Å. The ring brightness increases longward of 1600 Å, consistent with the decreasing absorption due to water ice (Wagener and Caldwell, 1988). The ultraviolet ring spectrum is quite similar to that of Saturn's retrograde moon, Phoebe, also seen close up for the first time by Cassini in 2004 (Esposito *et al.*, 2005). Similar to the Voyager observations, the A ring is the brightest in the ultraviolet spectrum, the C ring is the darkest, and the B ring is intermediate. The A ring spectrum is most similar to Phoebe, while the other rings are darker at wavelengths 1700–1900 Å, indicating a lower fraction of water ice. In the A ring, the ice fraction increases outward to a maximum at the outer edge. This large-scale variation is consistent with initially pure ice which has suffered meteoritic bombardment over the age of the solar system.

Infrared spectra from the Cassini Visual and Infrared Mapping Spectrometer (VIMS) tell a similar story. From the infrared absorption features of water ice,

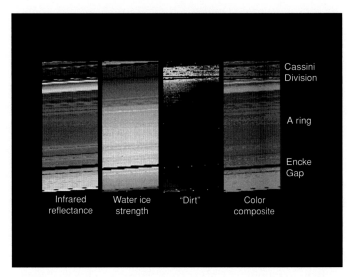

Figure 4.8 Spectra from the Cassini Visual and Infrared Mapping Spectrometer (VIMS) show evidence for an unidentified material called "dirt." The infrared reflectance image shows the brightness of scattered light transmitted through the rings. As Cassini observed the dark side of Saturn's rings just after entering orbit around Saturn, the empty and densest portions of the rings appear dark, while intermediate regions that scatter light toward the spectrometer appear brighter. The water ice strength image shows the amount of water in the rings, which peaks in the outer part of the ring A (bottom of panel). The "dirt" image shows that the dirty material is most abundant in the thinnest part of the rings: in the Cassini Division, in the Encke Gap, and in the other small gaps. In this image, multiple measurements of the same radial distance from Saturn are repeatedly measured; this gives the shallow diagonal banding. (NASA/JPL PIA 06350 "Dirty Rotten" Rings, 2 July 2004; courtesy NASA/JPL/University of Arizona.) For color version, see Plates section.

VIMS infers the amount of water in the rings, which peaks in the A ring. VIMS can also measure contaminants by their spectral signatures. An unidentified constituent is called "dirt" by Roger Clark of the US Geological Survey. This "dirt" is most abundant in the thinnest parts of the rings, and could also naturally be explained by the same process of meteoritic pollution more strongly affecting the less dense rings. The measured abundances from the VIMS data are shown in Figure 4.8.

   The reddish color of the rings has been ascribed to various organic materials, and more recently to exceedingly small iron materials (nano-iron or nano-hematite), mixed with the dominant water ice. In addition, regionally varying amounts of a gray absorber, such as carbon, can help explain the reflectivity; see a recent review by Cuzzi *et al.* (2009). Some spectra of Saturn's rings are similar to spectra of locations on Iapetus, which have strong evidence for mixtures of ice with nano-iron and nano-hematite (Clark *et al.*, 2013a, b).

Figure 4.9 Varying temperatures in Saturn's rings measured by the Cassini Composite Infrared Spectrometer (CIRS). Opaque regions, like the outer A ring (far right) and the middle B ring, are cooler, while the Cassini Division and C ring (innermost) are relatively warmer. (NASA/JPL PIA 06425 Saturn's Rings, Cold and Colder, 2 September 2004; courtesy NASA/JPL/GSFC/Ames.)

Observations by Cassini's Composite Infrared Spectrometer (CIRS) detect the variations in the temperature across the rings. Figure 4.9 depicts the varying temperatures. Temperatures range from about 70 K in the B ring to 100 K in the C ring and the Cassini Division. The opaque parts of the rings are colder, while the more transparent rings are warmer. Observations of the unlit side of the rings visible during closest approach show colder temperatures, consistent with ring particles rotating slowly. If the particles are rapidly rotating, their previously sunlit faces would come into view quickly and show a smaller temperature differential than Cassini observed (Flasar *et al.*, 2005).

4.6 Are rings gases or liquids?

It is clear that the nature of rings is that they are formed of solid particles. Each is an individual satellite of the planet. The total number is undeniably large. In order to investigate the behavior and history of rings, we must somehow reduce the sheer number of objects we consider. Two approaches have proven fruitful. The first is to consider each ring particle like a molecule of a gas. In this *kinetic theory* approach, a distribution function tells us the likelihood of finding a ring particle at a given location with a given speed. The alternative is to treat the rings like a fluid, the so-called "hydrodynamic approach." In this approach, the fluid has various properties like density, pressure, temperature, and velocity, which we identify with the average characteristics of rings. Furthermore, for the fluid analogy, we can define transport coefficients like viscosity and conductivity to explain how the ring evolves.

Each of these approaches has its advantages and disadvantages, which tells us that rings are more complicated than we might first imagine, and that both approaches are some form of idealization. As we see what works (and what doesn't) in explaining ring phenomena, we gain insights into the true nature of these beautiful structures. I will give some general statements of the long-term evolution, interactions, and rates of the various processes in planetary rings and proceed to list the master equations in each approach.

The theoretical explanation of rings has attracted the attention of physicists like Maxwell and Harold Jeffreys, and, more recently, planetary scientists and astronomers like Goldreich and Shu. A particular concern has been the stability and thus the history of the observed rings. Here, we follow the exposition of Stewart *et al.* (1984). A planetary ring consists of small particles that follow nearly circular orbits like those described by Johannes Kepler in explaining the Sun-centered solar system. Their orbital speed is given by Kepler's law,

$$\Omega = \sqrt{\frac{GM}{r^3}} \qquad (4.5)$$

where M is the central planet mass, G is the gravitational constant, r the distance from the center, and Ω is the angular rotation rate (say, degrees or radians per second). The actual rotational velocity is then given by:

$$v_{\mathrm{cir}} = \Omega r = \sqrt{\frac{GM}{r}} \qquad (4.6)$$

For real rings, these orbital velocities are in $\mathrm{km\,s^{-1}}$: even small ring particles would thus potentially be deadly to a spacecraft like Cassini. For this reason, spacecraft scrupulously avoid intersecting with rings. We recall from Kepler's laws that the innermost satellites revolve most rapidly. The same is true for rings – the inner parts move faster. Across the ring, the gradient of this velocity, the *Kepler shear*, has important consequences for the ring's history and evolution.

The second immediate effect is clear from the fact that the ring orbits are not perfectly circular. Ring particles thus collide frequently, causing the ring system to quickly flatten in the direction perpendicular to the mean orbital motion. For all ring systems, this plane is quite close to, although not necessarily identical to, the equatorial plane. If the local thickness of this plane is several times the diameter of the larger particles, more subtle effects occur. For a ring that is optically thick (that is, opaque), like Saturn's main rings, collisions occur about twice each orbit – this is a collision every few hours. Even the particles in the more transparent and ethereal rings would experience millions, if not billions, of collisions over the age of the solar system.

If we define the optical depth (the quantification of the ability of a ring to block light; see Chapter 15) of a ring of equal-sized particles as

$$\tau = \pi a^2 \sigma / m \tag{4.7}$$

where a and m are the radius and the mass of an individual particle and σ is the surface mass density (say, in $g\,cm^{-2}$) of the disk, then the collision frequency is estimated as 2τ each orbit, since a particle will cross the ring plane twice. This yields a collision frequency $v_c = \tau\Omega/\pi$. For an optically thick ($\tau \geq 1$) ring like Saturn's B ring, collisions occur every few hours.

This rapid collision rate explains why each ring is a nearly flat disk. If we start with a set of particle orbits on eccentric and mutually inclined orbits (say, the fragments of a small, shattered moon), collisions between particles dissipate energy: they are not perfectly elastic. At the same time, in these collisions, the laws of physics assure that the overall angular momentum of the ensemble is conserved. A particle in orbit cannot change its angular momentum without changing its mean distance from the planet. Thus, the relative velocity is damped out, and the disk flattens after only a few collisions to a set of nearly *coplanar*, circular orbits. If two ring particles collide, the inner one is near the *apoapse* of its (slightly) eccentric orbit, while the outer is closer to *periapse*. From Kepler's laws, we know the former is moving slower and the latter faster than its mean circular velocity. Afterwards, the loss of energy in the collision reduces their relative velocity and thus the eccentricity of the final orbit. A similar situation applies for two intersecting inclined orbits: in both cases, the ring evolves toward a (mostly) coplanar disk with (nearly) circular orbits.

After achieving this flattened state, the disk of the planetary ring evolves, but more slowly. The relative motion may now be dominated by the Kepler shear, the radial variation in orbital speed. The outermost collider is thus typically moving more slowly. The system continues to lose energy in collisions and to conserve momentum. Because the immediate result of the collision is toward equalizing the relative velocity at the intersection, the outer particle tends to be speeded up while the inner slows down. As these particles are nearly on circular orbits, the velocity changes lead directly to changes in the angular momentum about the central body, which in turn lead to a slightly different (mostly circular) orbit after the collision.

The inner ring particle loses angular momentum and falls to a lower (and thus by Kepler's laws, a faster) orbit, while the opposite is true of the outer particle. The result is that the two particles diverge. The overall evolution of the ring reflects this: it spreads. The net angular momentum transfer is outward, while the mass of the ring is gradually transferred inward. If there are no barriers to this inward diffusion, the ring particles will eventually reach the planet's atmosphere, each to burn up as a meteor. Donald Lynden-Bell and James Pringle (1974) found exactly the same result for *accretion disks* around black holes. This slow radial spreading means that collisions never disappear: the ring never reaches the perfectly flat, circular, and collisionless state. Instead, it reaches a quasi-stationary state a few particles thick.

4.7 Evolutionary rate

How fast is this spreading? Consider the *mean free path* λ (average radial distance between collisions). For thick rings, this is just the average random speed c (let us break up the total velocity as follows: $v = v_{\text{cir}} + c$, where $c \ll v_{\text{cir}}$) multiplied by the time between collisions: $\lambda = c/(\Omega\tau)$. For very thin rings, the mean free path is given by $\lambda = c/\Omega$, since the time between collisions is more like the orbital period. Alan Cook and Fred Franklin (1964) included both limiting values in their prescription (which was also adopted by Goldreich and Tremaine, 1978):

$$\lambda^2 = \frac{c^2}{\Omega^2} \frac{1}{1 + \tau^2} \tag{4.8}$$

The behavior of any individual particle experiencing repeated collisions can be seen as an aspect of a simple *random walk* (see Chapter 9) with the step size in radius given by λ. Let $\Delta r = n\lambda$. For a random walk, it takes on the average n^2 steps to reach a distance $n\lambda$ from the origin. Thus, the time for a typical particle to diffuse a distance Δr is n^2 steps each of duration $\Delta t = 1/(\Omega\tau)$.

Putting it all together, we have, for the total time for a particle to diffuse a distance Δr:

$$T \approx \frac{n^2}{\Omega\tau} = \left(\frac{\Delta r}{\lambda}\right)^2 \frac{1}{\Omega\tau} = \left(\frac{\Delta r}{c}\right)^2 \Omega \frac{1 + \tau^2}{\tau} \tag{4.9}$$

This process may alternatively be considered as diffusion in a fluid, in which case the viscosity is given by $v = \lambda^2 \Omega\tau$. For a fluid, we have $T = \Delta r^2/v$, the same as above. Thus, at this level of accuracy, both the gas and fluid approximations for the ring give the same timescale for the process of ring spreading.

4.8 Energy transfer

The Kepler shear causes the collisions and the slow spreading of a planetary ring as explained above. At each collision, some energy is lost from the general Keplerian motion and converted to the particular motion of individual particles. If local processes dominate, then these two processes must exactly balance. In principle, this could allow us to determine an equilibrium velocity for the ring particles, but we are limited in calculating this for real rings by our incomplete knowledge of the distributions of size and velocity for the ring particles, along with the possibility of fragmentation or accretion when particles collide. Furthermore, we would need to know the exact value of the elasticity of every impact in each of the possible collisions. This measure of elasticity is the coefficient of restitution (4.1) and may well vary with size and velocity. Nonetheless, it is clear that the elasticity declines with increasing impact velocity, and thus some equilibrium is possible (Stewart *et al.*, 1984). Thus, as Kepler shear introduces collisions and pumps up the random velocity c, more energy is lost in each subsequent collision until the equilibrium is achieved. Conversely, if c falls below its equilibrium value, energy losses decline from impacts and the random velocity increases to re-establish equilibrium.

4.9 Planetary rings as a gas: kinetic theory approach

This approach is based on describing the distribution of ring particles by a distribution function $f(m, r, v, t)$, which is the *probability density* of finding a particle of mass m at position r with velocity v at time t. We note that f is the density of particles in a hyper-dimensional *phase space*. In terms of f, the evolution of the ring system is described by the *Boltzmann equation* from the kinetic theory of gases (Stewart *et al.*, 1984):

$$\left(\frac{\partial}{\partial t} + v_i \frac{\partial}{\partial x_i} - \frac{\partial U}{\partial x_i} \frac{\partial}{\partial v_i}\right) f(m) = \sum_j C\left[f(m),\ f(m_j)\right] \qquad (4.10)$$

where x, v are the position and velocity and the indices represent *orthonormal* coordinates, and $U(x)$ is the gravitational potential due to the planet and its satellites (not included in f) and the other ring particles. The right-hand side represents the rate of change of f caused by collisions with particles of mass m_j.

   This equation is purely formal and the complexity of the *collision integral* prevents any exact solution. We can replace the original equation, (4.10), by three sets of *moment equations* (multiply by 1, v_i, and $v_i v_j$ and integrate over all velocities; see Goldreich and Tremaine (1978)). The result is entirely analogous to the kinetic theory of gases. Once again, the *difficult* physics is now within the three collision integrals on the right-hand side. The solution to these moment equations entails the twofold problem of (1) including the appropriate physics in the collision integrals and (2) calculating the solution. Naturally, these two objectives are in tension; many clever solutions exist but none that fully meets the requirements of a complete answer.

4.10 Planetary rings as a fluid: hydrodynamics

Jack Lissauer and Glen Stewart (1993) note that although the kinetic theory treatment is rigorously correct, its complexity makes exact analytic solutions unattainable. Numerous approximations and simplifications are required to deduce the basic results. The mathematical approach can obscure the central issue of interpretation of the essential physical processes that determine the structure and evolution of the rings. However, these complexities may be reduced by the use of a fluid dynamical approximation.

In this approach, the *pressure tensor* is divided into two parts: an *isotropic* part, which we interpret as the normal pressure; and a non-isotropic part, called the *stress tensor*, which is the product of the local transport coefficient (viscosity) and the global shear. Through this approximation, the challenge of calculating collision integrals is replaced by the task of finding prescriptions for the transport coefficients.

As an example, Jürgen Schmidt *et al.* (2001) write the following three equations for the surface density σ, the velocity $\vec{u}$, and the temperature T of a thin disk:

$$(\partial_t + \vec{u} \cdot \vec{\nabla})\sigma = -\sigma \vec{\nabla} \cdot \vec{u} \qquad \text{(continuity)} \qquad (4.11)$$

$$\sigma(\partial_t + \vec{u} \cdot \vec{\nabla})\vec{u} = \sigma \vec{F} - \sigma \vec{\nabla}\Phi - \vec{\nabla} \cdot \hat{P} \quad \text{(momentum)} \qquad (4.12)$$

$$3/2\sigma(\partial_t + \vec{u} \cdot \vec{\nabla})T = -\hat{P} : \vec{\nabla}\vec{u} - \vec{\nabla} \cdot \vec{q} - \Gamma \quad \text{(energy)} \qquad (4.13)$$

The term Γ is the dissipation of energy of the motion due to inelastic collisions. The gravitational field of the planet is $\vec{F}$; the disk's mean self-gravity potential is Φ; and the heat flux vector is $\vec{q}$. The momentum equation is the familiar Navier–Stokes equation of fluid dynamics, and the energy equation is the heat-flow equation of *viscous* hydrodynamics. These same equations apply to other systems approximated as fluids, for example, flow of granular material. The key transport coefficients are the kinematic viscosity, the bulk viscosity, and the heat conductivity, which can be directly connected to the collision integrals of the kinetic theory (above). Mark Lewis and Stewart (2000) have noted that their detailed simulations of planetary rings perturbed by nearby satellites show that the fluid approximation breaks down there; Schmidt *et al.* (2001) note that the transport coefficients neglect the anisotropy of the velocity distribution and that a better description is provided by the kinetic theory. These comments show that the fluid approximation must be used with care; nonetheless, it allows us to connect the behavior of the rings with that of more (or less) familiar fluids.

4.11 Summary

Individual ring particles are most likely aggregated into "rubble piles" that repeatedly grow and are disrupted by tides, collisions, and impacts. The nature of their collisions determines the energy balance, the transfer of momentum by viscosity, and particle size distribution. The largest disruptions are visible to spacecraft as individual events. Some disruptive phenomena may have been observed by Cassini; see Chapter 7.

A formal description of ring physics can be written with a small number of equations. The challenge is to incorporate the key physics into the terms of these equations, which then must be solved. Although the kinetic theory description is more inclusive, solutions are few. The hydrodynamic approach can yield results that mesh with our understanding of fluids, at the cost of making some assumptions. In either case, we must be aware of the simplifications and approximations used. These results allow us to calculate the detailed history of planetary rings and simulate the phenomena seen in them. We can determine under which conditions the rings are stable and under which they are not. Thus, the idea that planetary rings are a gas where the small deviations from circular orbits are like the gas molecules' random motions, or the idea that the rings are a slowly diffusing fluid, will be useful to us in the chapters that follow. The multiple approaches serve to illuminate the true dynamics. The rings of a planet, thus, are both a gas and a liquid!

5

Large-scale ring evolution

The first modern step toward understanding planetary rings was made just before the discovery of the Uranus rings in 1977 through the study of simple collisional systems unperturbed by nearby objects. The collisional dynamics of a differentially rotating disk of particles have been studied extensively; see, for example, Lynden-Bell and Pringle (1974), Brahic (1975, 1977), Goldreich and Tremaine (1978), and Lin and Papaloizou (1979). The main results can be summarized (Brahic, 1977): after a very fast flattening within a time of the order of a few tens of collisions per particle, the system reaches a quasi-equilibrium state in which the thickness of the newly formed disk is finite (i.e., the centers of the particles do not lie in the same plane) and in which collisions still occur. Under the combined effect of *differential rotation* and of inelastic collisions, the disk spreads very slowly; particles move both inwards and outwards carrying some angular momentum while conserving the total angular momentum of the system. In the absence of external confining forces, the spreading time is of the order of the time it takes particles to execute a random walk of a distance equal to the ring width (see Chapter 4). During particle collisions, part of the relative velocity arising from differential rotation is transformed into vertical and radial motion. A steady state is established between this kinetic energy received by the system from differential rotation and that drained away through inelastic collisions, so that the kinetic energy that is continually lost by individual particles is obtained at the expense of potential energy from the bodies moving inwards and outwards. Because angular momentum is conserved, the energy lost by the inward motion of a portion of the particles is larger than the energy gained by the remainder that are moving outwards: to spread, the disk gives up a small amount of its total energy.

5.1 Particle dynamics

The first analytical studies of ring dynamics solved the Boltzmann equation (see 4.10) with a specified collision integral on the right side. In order to solve this equation, Goldreich and Tremaine (1978, 1982), Suguru Araki and Scott Tremaine (1986), Frank Shu and Glen Stewart (1985), and K. A. Hameen-Anttila (1978) have introduced simplifications to the complete Boltzmann equation based on several assumptions. An important question is to relate the numerical results to real rings. Rather than trying to reproduce any particular observed feature, these studies have served to investigate the important physical mechanisms. More modern research has used *N-body simulations* to explicate the phenomena (see Chapter 8, and, e.g., Salo, 1995; Salo and Hänninen, 1998; Lewis and Stewart, 2000, 2009). Lewis and Stewart (2000) show that the ring dynamics are significantly more complex than a standard fluid model, owing to the actual sizes of particles and alignment of particle *epicycle* phases.

5.2 Mechanisms for ring confinement

A significant mechanism for confining rings has been quantitatively studied by Goldreich and Tremaine (1979, 1982). Through gravity, a nearby satellite will alter a ring particle's orbit, making it elliptical. This effect is especially pronounced where the period of the ring-particle orbit is in the ratio of small integers to the period of the perturbing satellite. Such a location is termed a resonance and labeled by the ratio of periods; that is, at the "Mimas $2:1$ resonance," a ring particle orbits Saturn twice each time Mimas orbits once. The overall effect of the resonant *perturbations* on a population of particles in a disk is to increase the density of particles in some places and to decrease it in other places; this pattern is static in the satellite's frame but moves through the disk and thereby generates spiral density waves like those in spiral galaxies (see Figure 5.1). If the perturbing satellite is exterior to the ring, this wave moves outwards from the resonance, carrying negative energy and negative angular momentum (see Figure 5.1). An isolated ring particle would not suffer any systematic drift after the encounter with the satellite (although its subsequent motion is complicated). Because rings are dense enough to have numerous collisions between particles, the particles involved in such collisions gradually move inwards toward the planet. In other words, encounters with a satellite increase radial motions, while collisions circularize the orbits. A balance is achieved in which the orbits of the particles accommodate to the perturbing influence (Brophy *et al.*, 1992). The net effect is thus a repulsion of the orbits due to the gravitational attraction between the ring and moon (see Greenberg, 1983). Cuzzi (1998) compares this shepherding to an overly protective parent who ends up repelling rather than attracting his children (see Figure 5.2). The exterior satellite removes angular momentum from the ring and adds energy to nearby particles, thus transferring angular momentum from the ring to its orbit (see, e.g., Lin and Papaloizou, 1979). If the angular momentum luminosity carried off by

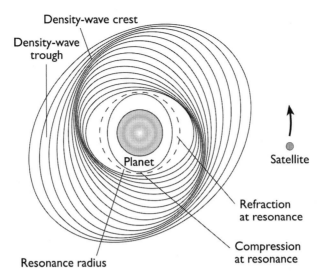

Density-wave crest
Density-wave trough
Planet
Resonance radius
Satellite
Refraction at resonance
Compression at resonance

Figure 5.1 The creation of a spiral density wave begins when a two-armed spiral wave is excited at the 2 : 1 resonance (shown dashed) with the exterior moon. The ovals represent particle paths as seen in the frame rotating with the orbiting satellite. Their long axes become less and less well aligned with the satellite's direction at greater distances from the resonances. The clustering that occurs in the orbital paths induces coherent oscillations in neighboring particles as they drift past by Keplerian shear. Actual spiral waves are much more tightly wrapped than shown here. (Burns (1999), from *The New Solar System*, 4th edn, edited by J. Kelly Beatty, Carolyn Collins Petersen, and Andrew Chaikin. Copyright © 1999 Sky Publishing Corp. Reprinted with permission.)

this wave exceeds that carried by the particles diffusing across the resonance, the wave truncates the ring and an edge is formed (Goldreich and Tremaine, 1979). An example is provided by the Uranus system, where known satellites (e.g., Cordelia and Ophelia on either side of the ε ring) as well as small undetected satellites (for the other rings) on each side of a narrow ring can thus constrain its edges and prevent the ring from spreading. Elliptical and inclined rings can also be generated by such a confining mechanism (Goldreich and Tremaine, 1982).

The precise mechanism for shepherding has not been completely determined (Borderies *et al.*, 1989; Brophy *et al.*, 1990; Lewis and Stewart, 2000; Lewis, 2001) but can be qualitatively understood as follows. Satellites exert torques on the boundary of a ring located at a low-order resonance. For the simplest case of a circular ring and a circular satellite orbit, the strongest torques occur where the ratio of the satellite orbit period to the ring-particle period equals $m/(m + 1)$ with m an integer. The torque is of the order of:

$$T_m = \pm m^2 \left(G^2 m_s^2 \sigma\right)/(\Omega^2 r^2) \tag{5.1}$$

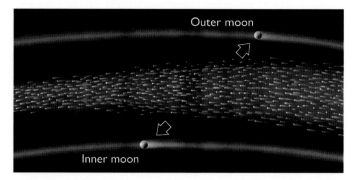

Figure 5.2 The Goldreich–Tremaine model for constraining narrow rings employs "shepherd satellites" that force a group of particles to travel along narrow paths. Moving slower than the ring interior to it, the outer satellite attracts particles going by. This attraction is slightly greater after particles have passed the satellite, because when slipping by it their paths were pulled somewhat closer (not shown) to the shepherding object. This extra force causes particles to lose energy and "fall" closer to the planet. Conversely, the faster-moving inner satellite adds energy to nearby particles and kicks them into higher orbits. Together, these forces herd the particles into a narrow ring. The same process can explain how an embedded satellite could open a gap in a disk, by spreading the system apart. (Figure by Don Davis, from *The New Solar System*, 4th edn, edited by J. Kelly Beatty, Carolyn Collins Petersen, and Andrew Chaikin. Copyright © 1999 Sky Publishing Corp. Reprinted with permission.)

(Goldreich and Tremaine, 1980). G, Ω, m_s, r, and σ are, respectively, the gravitational constant, the orbital frequency, the satellite mass, the orbital radius, and the surface mass density evaluated at the resonance location. If the spacing between neighboring resonances from a nearby satellite is very small, the widths of individual resonances can be greater than the separations, so that resonances overlap. Thus, it is useful to sum the discrete resonance torques and to define the total torque on a narrow ringlet of width Δr:

$$T \sim \pm \left(G^2 m_s^2 \sigma r \, \Delta r \right) / (\Omega^2 x^4) \tag{5.2}$$

where x is the separation between the satellite and the ringlet ($r \gg x \gg \Delta r$) (Goldreich and Tremaine, 1980).

The presence of dissipation is essential, since the torque would vanish without it. In that case, the nearby moon has no long-term effect on the ring. The exact nature of the dissipation does not affect the expression of the torque (Greenberg, 1983; Meyer-Vernet and Sicardy, 1984), as suggested by the fact that the torque expression does not explicitly contain any dependence on dissipation. Particle collisions are evidently the main source of dissipation. The physics involving many colliding particles sets the study of ring dynamics apart from classical celestial dynamics, in which no collisions are assumed, and fluid dynamics, in which more frequent collisions are known to occur than in planetary rings (where the mean free path of an individual particle may be much longer than in the fluid case; see equation 4.8). Further, because each ring particle is on an individual orbit around the planet, unlimited separations are impossible.

Note that the resonance torque would not act on isolated test particles. For example, because of the rarity of collisions in the asteroid belt, this mechanism is not responsible for the formation of the empty regions in the asteroid belt, known as the Kirkwood gaps. Instead, chaotic motion of asteroids excited by Jupiter depopulates this region and may bring meteorites to Earth.

The rate of transfer of angular momentum can also be calculated by a perturbative approach without reference to individual resonances. The gravitational interaction of a ring particle with a satellite occurs primarily close to encounter. Their relative motion is not perfectly symmetric around encounter, so that the tangential component of the relative velocity of the particle with respect to the satellite is reduced, and thus angular momentum is exchanged with the net result that the ring experiences a torque (Lin and Papaloizou, 1979). Furthermore, a particle initially moving on a circular orbit acquires a radial velocity and thereafter moves on a Keplerian ellipse. In a frame corotating with the perturbing satellite, all particles initially moving in circular orbits must follow similar paths after encounters. Thus, each perturbing satellite generates a standing wave. In the inertial frame, each particle moves on an independent Keplerian ellipse, but the pericenters of these elliptical orbits and the phases of the particles create a sinusoidal wave that moves through the ring with the angular velocity of the perturbing satellite. For a more complete discussion, see Shu (1984). Numerous such density waves have been observed in the rings of Saturn (see below) and perhaps also in those of Uranus (Horn *et al.*, 1988). The damping of these waves by collisions can result in a net exchange of angular momentum between the satellite and the ring particles. This phenomenon is similar to the dynamical friction studied in stellar dynamics.

The discovery of 10 small satellites inside the orbit of Miranda during the Voyager 2 Uranus encounter has given a remarkable opportunity to test the predictions for shepherding of the Uranus rings. Porco and Goldreich (1987) showed that Cordelia and Ophelia are the inner and outer shepherds for the ε ring, and that Cordelia is an outer shepherd for the γ ring. They have demonstrated that these satellites are capable of confining the ε ring if the mass and the thickness of the ring are low enough. The outer edges of the δ and the γ rings can also be confined by these satellites, but the drag due to the planet's extended neutral hydrogen *exosphere* would require larger shepherd moons and thus poses a severe problem for the shepherding of the α and β rings (see Chapter 10 and French *et al.*, 1991).

5.3 Summary

The large-scale evolution of rings is determined by the combined effects of collisions and differential rotation. Approximating the ring system as a gas of interacting particles captures aspects of ring dynamics better than a fluid model but neither fully reproduces the rings' behavior. Rings can be confined by the gravity of small nearby moons, called "shepherds." Observations of the Uranian ε ring confirm this theoretical prediction.

6

Moons confine and sculpt rings

The expectation of smooth, featureless, gradually spreading rings (see Chapter 5) is nowhere observed in nature. Instead, the rings show structure at all scales, radial, vertical, and azimuthal. Esposito *et al.* (1983a) used time-series analysis of the Voyager 2 ring stellar occultation to show fluctuations at all length scales down to the *noise* in their data, certainly well below 1 km. Cassini star occultations now show structures just a few meters across. Saturn's rings are known to be warped and bent. The rings of every planet show numerous azimuthal asymmetries. The probable cause for much of this structure is small moons near and within the rings.

6.1 Resonances

The ability of such small satellites to create observable phenomena lies in the effectiveness of resonance forcing. Even though the mass of the satellites and their gravity is much less than the central planet's, repeated perturbations at a natural frequency of the ring particles can build up substantial effects. The small signal is amplified in the same way that a playground swing can be pumped at its natural frequency or annoying feedback can occur at a rock concert. To quantify this idea, we need to know both the natural frequencies of the rings and the frequencies of the moon's *gravitational perturbation*. The forcing frequency is obviously related to the moon's motion. For a moon on a circular, uninclined orbit around a spherical planet, the only frequency is the mean motion, $n_s = (GM/r_s^3)^{1/2}$. In the more general case, the elliptical motion can be considered as a small centered ellipse about the guiding center of the circular orbit, with the *epicyclic frequency* κ_s identical to n_s. For an orbit with small inclination, we can use the same approximation of an equatorial orbit plus a small vertical oscillation. For an *oblate* planet with an *axisymmetric* potential (expressed in terms of the *zonal harmonics* J_2, J_4, etc.), the orbit rotates, yielding three distinct frequencies $\mu_s > n_s > \kappa_s$. These are, respectively, the vertical frequency, the mean motion, and the epicyclic frequency. The inequalities express the fact that the orbit line of *apsides* precesses, while the *node* regresses. Thus, the forcing potential felt by a ring particle is composed of these natural frequencies of the perturbing moon and their harmonics. A Fourier series analysis can decompose the gravitational field of the moon into its components, the strongest of these located at some frequency where the argument (representing the combination of the various natural frequency harmonics) is constant. An observer rotating with that angular rate would see a repeating pattern. Thus, the pattern speed, Ω_p, satisfies:

$$W_f = m\Omega_p = mn_s + k\kappa_s + p\mu_s \tag{6.1}$$

where W_f is the forcing frequency, and m, k, and p are integers. The integer m gives the multiplicity of the instantaneous pattern. For the simplest case of a circular, equatorial orbit around a *point-mass planet*, the forcing frequencies are just the multiples of the moon's mean motion. Thus, resonances occur at locations where the natural frequency of the particle's motion is the same as the forcing felt in a frame rotating at the particle's orbital motion.

The natural frequencies of a ring particle (for definiteness, inside the orbit of the perturbing moon) are just the values of n, μ and κ appropriate for that particle's orbit. Thus, we can identify three possible types of resonance, where the forcing frequency matches

(1) the particle's mean motion: "*corotation* resonance"

$$W_f - mn = 0 \tag{6.2}$$

(2) the epicyclic frequency: *Lindblad resonance*

$$W_f - mn = \pm\kappa \tag{6.3}$$

(3) the vertical frequency: *vertical resonance*

$$W_f - mn = \pm\mu \tag{6.4}$$

The upper signs are for the perturbing moon inside the resonance location, the lower for the perturber outward from the resonance.

6.2 Pendulum model

Murray and Dermott (1999) show that the motion of a resonant body (neglecting some small terms) is:

$$\ddot{\Phi} = -\omega_0^2 \sin \Phi \tag{6.5}$$

where Φ is the resonant argument that measures the angular distance from perfect resonance ($\Phi = 0$). This means that the *libration* of a ring particle around resonance is similar to pendulum motion. Particularly, for small Φ (quite close to resonance), we can make the approximation $\sin \Phi \approx \Phi$, so that we have:

$$\ddot{\Phi} = -\omega_0^2 \, \Phi \tag{6.6}$$

the motion of a simple *harmonic oscillator*. The pendulum has both circulating and librating solutions, with the particular solution depending on the initial conditions. If the initial energy is large, then the resonant motion of the argument is *unbounded*, which yields circulation of the angle Φ. This corresponds to a 360° motion of a pendulum around its point of suspension. If the initial energy is small, Φ is bounded: the resonating body oscillates or librates about $\Phi = 0$. These two cases are divided by the *separatrix*, which represents a solution with infinite period.

The point at $\Phi = 0$ is an elliptic fixed point and stable; the point at $\Phi = \pm\pi$ is a hyperbolic fixed point and unstable. These ideas can assist in estimating the variation in orbital parameters caused by individual resonances. Estimating the libration in *semi-major axis* (equivalent to mean motion, n) gives a predicted width for ring features created by resonances. Particles nearest the resonance will have the largest amplitudes: the *forced* eccentricity falls off inversely as the distance from the resonance in the simplest linear theory. Beyond the hyperbolic fixed point, resonant effects will be minimal.

The above classification scheme covers the strongest resonances in planetary rings. The notation $(m+k+p):m-1$ is used to identify the individual resonances, since $\mu \approx n \approx \kappa$, giving the ratio (for moons outside the resonance) of the mean motion of the particle to the moon:

$$\frac{n}{n_s} \approx \frac{m+k+p}{m-1} \tag{6.7}$$

Thus, the resonance at 88 705 km from Saturn's center where $m=2$, $k=0$, $p=1$ is called the Mimas 3:1 inner vertical resonance (IVR). A Cassini image of the *bending wave* excited at this resonance is shown in Figure 6.1. The location of the Mimas $m=2$, $k=1$, $p=0$ resonance (inner Lindblad resonance, ILR) also excites a wave. The positions and strengths for strongest resonances in Saturn's rings are shown in Figure 6.2. The majority of these resonances are found in the outer part of Saturn's A ring, since it is the ring closest to the perturbing moons. The resonance strength can be written to lowest order

$$T \sim M_s e^k \sin^p i \tag{6.8}$$

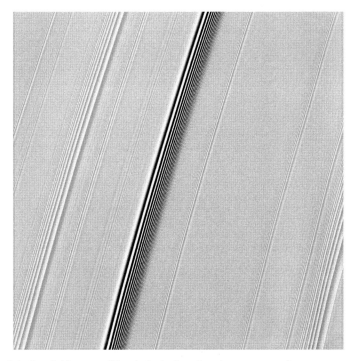

Figure 6.1 Cassini image of Saturn's A ring showing two prominent wave patterns. The feature on the right is the Mimas 5 : 3 bending wave; its contrast is high because the tilt of the local ring plane due to the wave was comparable to the solar elevation angle when the image was taken. The Mimas 5 : 3 density wave is on the left. The separation between the locations of the two waves results from the non-closure of orbits caused by Saturn's oblateness. The other linear features in the images are density waves excited by the moons Pandora and Prometheus. Saturn is off to the right. (PIA12545; courtesy NASA/JPL/Space Science Institute.)

where M_s, e, and i are the mass, eccentricity, and inclination of the perturbing moon. Since e and i are small quantities for most moons, the strongest resonances have $k = 0$, $p = 1$ or vice versa.

6.3 Shepherding

One of the surprising developments of studies of planetary rings is that moons and rings actually repel each other through their gravitational interactions at resonance.

This is another example of the ability of a moon to repel ring material, called shepherding, because the moon acts like a "shepherd" to constrain the "flock" of ring particles. This phenomenon arises when ring particles are perturbed into eccentric orbits by their encounter with the moon. Collisions between ring particles tend to circularize the orbits before the next encounter, resulting in a net repulsion.

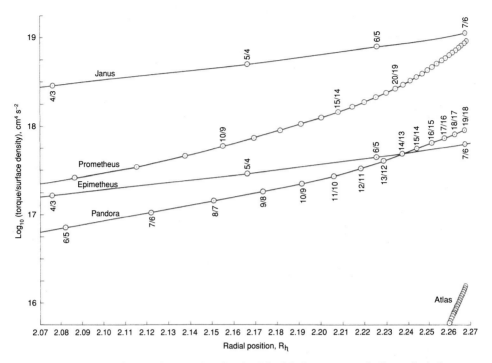

Figure 6.2 Locations and strengths of major Lindblad resonances in Saturn's A ring for Saturn's five closest "ring moons," whose orbits lie exterior to the main ring system: Janus, Epimetheus, Pandora, Prometheus, and Atlas. The moons that orbit closer to the A ring have more closely spaced resonances with strength increasing outward more rapidly. (Adapted from Lissauer and Cuzzi (1982), from de Pater and Lissauer (2001); reprinted with the permission of Cambridge University Press.)

Since the strongest effects occur when the moon is near the ring particle, a good approximation is to assume a single impulse at conjunction. If we assume initially circular orbits for both the ring particle and the moon, the new eccentricity of the ring particle is:

$$e \approx 2.24 \frac{M_{\rm s}}{M_{\rm p}} \left(\frac{a}{\Delta a} \right)^2 \qquad (6.9)$$

where $M_{\rm s}$ and $M_{\rm p}$ are the moon and planet mass, a is the particle *semi-major axis*, and Δa the separation of the two orbits (Julian and Toomre, 1966; see an approximate derivation in Murray and Dermott, 1999). After the impulse, the ring particle is on a new eccentric orbit. The particles execute epicyclic motion after the encounter with radial amplitude ae. Since the distance along the edge of a ring of particles just measures the time since encounter, a ring of particles perturbed by a

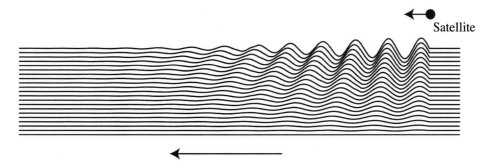

Figure 6.3 A schematic diagram showing how the variation of the amplitude and wavelength with separation from a satellite leads to the formation of a wake in adjacent ring material upstream from an exterior satellite. The lower arrow indicates the direction of motion of the particles with respect to the satellite. (From Murray and Dermott (1999); reprinted with the permission of Cambridge University Press.)

nearby satellite will have a sinusoidal edge expressing the epicyclic motion, with wavelength

$$l(\text{edge}) = 3\pi \ \Delta a \frac{n}{\kappa} \approx 3\pi \, \Delta a \qquad (6.10)$$

(see Figure 6.3) since the relative angular velocity is (by Kepler's third law)

$$u = \frac{3}{2}\frac{n}{a}\Delta a \qquad (6.11)$$

and the epicyclic period is

$$T = \frac{2\pi}{\kappa} \qquad (6.12)$$

In a time T, the ring particle moves downstream a distance $l(\text{edge})$:

$$l(\text{edge}) = uaT = 3\pi \, \Delta a \qquad (6.13)$$

Measuring the wavelength of the wake formed this way allows us to measure the distance Δa to the perturbing moon (e.g., Showalter *et al.*, 1986).

The small transfer of momentum is a consequence of a slight asymmetry in the encounter, so that the objects are slightly closer after their closest approach. Alternatively, we can consider the gravitational interaction to have an effect toward reducing the ring particle's and moon's relative velocity at encounter. This slows the inner object and speeds the outer one so that their orbits would diverge when circularized. If the particle orbits are not circularized before the next encounter, a reverse transfer can occur at future encounters so that the total momentum transfer over long times is net zero. A satellite on either side of the ring can hold it in

Figure 6.4 Cassini image of the moon Daphnis, which opens the Keeler Gap and creates scalloped edge waves on the gap edges, similar to the moon Pan's effects on the wider Encke Gap. Daphnis has a diameter of 7 km. (PIA06237; courtesy NASA/JPL/Space Science Institute)

place against its natural tendency to spread (see Figure 5.2). This explanation was initially developed by Goldreich and Tremaine (1979) to explain the surprising phenomenon of Uranus' narrow rings with sharp edges. Voyager photos in 1980 of Saturn's F ring seemed to provide a visual confirmation of their theory. However, a number of small difficulties and results of more detailed simulations by Lewis and Stewart (2000) imply that the theory presented here is still oversimplified.

We see that a moon pushes material away from it if it is large enough and close enough to overcome the natural spreading of the ring. Thus, a moon (or a big enough ring particle) could clear a gap in a ring. Spectacularly, the small moon Pan was discovered in Saturn's Encke Gap by Showalter (1991): his search was motivated by the wavy edges of the gap and the wake visible around it. Daphnis similarly clears the Keeler Gap in the outer A ring (see Figure 6.4).

6.4 Waves

Resonance with nearby moons has a second observable effect. In addition to clearing gaps and creating sharp edges, this resonant forcing also launches waves that propagate in the rings. More than 50 examples have been detected in Saturn's rings (Esposito *et al.*, 1983a; Tiscareno *et al.*, 2007) and perhaps in the Uranian ring (Horn *et al.*, 1988). The basic theory of spiral density waves, originally developed to explain spiral galaxies, was applied to Saturn's rings by Goldreich and Tremaine (1978). The effect of a resonance is to create an azimuthal variation in the gravitational potential in the rings with a number of maxima equal to the index m. Streamlines of the particle motion have an m-lobed pattern. Because of differential rotation due to Kepler's third law, this gives rise to a spiral pattern with m spiral arms. Each individual particle still has its own Keplerian elliptical orbit, but the superposition yields a spiral variation in particle number density. These horizontal density oscillations are termed *density waves*. Similarly, vertical corrugations arising from the inclinations of the particle orbits are spiral *bending waves*. These normal modes of the disk are excited at the resonance and propagate because the gravity of the ring particles themselves provides a restoring force. The linear theory (Shu, 1984; Tiscareno *et al.*, 2007) and non-linear theory (Shu *et al.*, 1985a, b) show some beautiful comparisons between planetary rings and galaxies.

In our discussion of resonance above, we noted that each ring particle could be considered a simple harmonic oscillator with three resonant frequencies. These individual oscillators can also communicate with each other through their own gravity: we may imagine them as (differentially rotating) pendula connected by strings. The self-gravity represented by the string is a restoring force that allows waves to propagate away from the resonance.

Distant from the resonance, we have a spiral pattern with radial wavelength

$$\lambda = \frac{4\pi^2 G\sigma}{m^2[W_f - n] - \kappa^2} \tag{6.14}$$

where G is the gravitational constant, σ is the surface mass density of the ring, W_f is the forcing frequency, and $n(r)$ and $\kappa(r)$ are the mean motion and epicyclic frequency of the ring particle at position r. For bending waves, replace κ by μ.

Approximating the ring-particle orbits as Kepler ellipses, $n(r) = \kappa(r) = \left(\frac{GM_p}{a^3}\right)^{1/2}$ for $m > 1$; and for $m = 1$ approximating their departure from Kepler orbits as just due to Saturn's *quadrupole* gravitational moment (J_2) we find:

$$\lambda(r) \approx 3.08 \left(\frac{r_L}{R_s}\right)^4 \frac{\sigma}{m-1} \frac{1}{r-r_L} \qquad m > 1 \tag{6.15}$$

$$\approx 54.1 \left(\frac{r_L}{R_s}\right)^6 \sigma \frac{1}{r-r_L} \qquad m = 1 \tag{6.16}$$

where λ, r, and r_L (resonance location) are measured in kilometers and σ in g cm^{-2} (e.g., de Pater and Lissauer, 2010). R_s is Saturn's equatorial radius; $R_s = 60\,330$ km. For bending waves, replace r_L by r_V, the radial location of the vertical resonance. Note that the predicted behavior of the wavelength is to decrease inversely as the distance from resonance. This particular dispersive behavior has been used to search for such waves by time-series analysis (Esposito *et al.*, 1983a; Spilker *et al.*, 2004; Tiscareno *et al.*, 2007; Baillie *et al.*, 2011).

The non-linear theory shows that density waves are not sinusoidal, but sharply peaked, as observed for most waves in Saturn's rings. Further, these sharp peaks are separated by shallow troughs (see Figure 6.5). Despite these differences in shape, the torque transferred by the waves is similar in both the linear and non-linear cases.

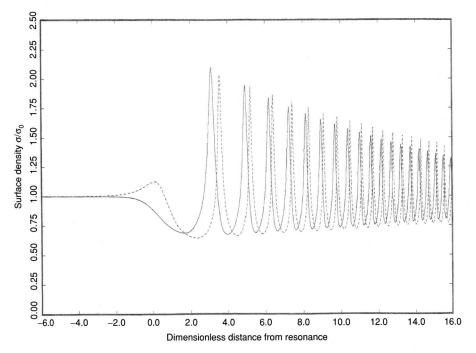

Figure 6.5 Theoretical surface density profile of a damped non-linear spiral density wave. The solid and dashed lines represent two profiles of the same wave plotted at different *azimuths*. (Adapted from Shu *et al.* (1985b) by de Pater and Lissauer (2001); reprinted with the permission of Cambridge University Press.)

This torque is balanced by that which the rings exert on the inner moons, through conservation of angular momentum. This causes those moons to recede from the rings at a rate that is very short compared to the age of the solar system: 10^6 to 10^8 years! This outstanding problem of the age and origin of the rings is discussed below (Chapter 10); unless another process counterbalances this evolution, the rings must be much younger than the solar system and perhaps recently created (Esposito, 1986).

6.5 Longitudinal confinement

Another unfulfilled expectation about planetary rings is that they would be azimuthally uniform. Since both the orbital rate (n) and *precession* rate ($n-\kappa$) decline with distance from a planet, it is difficult for a ring to maintain azimuthal structure: any such structure should be rapidly smeared out on the timescale of a small number of orbits by Keplerian shear. Naturally, the gravity of small moons could provide the forces to counteract this and maintain longitudinal structure. In the case of Neptune's Adams ring, the discontinuous arcs could persist against the Kepler shear through a combination of Lindblad and corotation resonances.

In the case of Neptune, ground-based observations of ring occultations showed blockage by rings only 10% of the time. Early on, it was easy to dismiss these few occurrences as observational errors, until two separate observers saw the same event (Hubbard *et al.*, 1986). Then it was realized that Neptune's rings must be discontinuous, as was spectacularly confirmed by Voyager observations in 1989. Lissauer (1985) proposed that the arcs were located at the Lagrange points of a small satellite: their libration would be limited in longitude, and thus provide *azimuthal confinement*. Additional satellites could provide the Lindblad resonances for radial confinement. A more economical solution was provided by Goldreich *et al.* (1986), in which a single satellite delivers both the radial and azimuthal confinement. A corotation resonance provides a number of equilibrium points, and a Lindblad resonance excites particle eccentricity (needed to provide energy, since the Lagrange points are energy maxima). Without this supply of energy, the dissipative collisions between particles would cause them to spread apart and eventually escape from the resonances into circulating orbits.

Some problems remain, even with the latest explanations for Neptune's ring arcs (see Chapter 13). It is clear that moons seen, and perhaps yet unseen, provide the forces to explain the longitudinal variations and their persistence.

6.6 Summary

The effect of small moons to confine and sculpt rings arises from resonances. The natural frequencies of ring particles orbiting an oblate planet can be compared to that of the pendulum. For small amplitudes, this means each ring particle behaves like a simple harmonic oscillator. These resonances can cause gaps, sharp edges, corrugations along the edges, and radial and longitudinal confinement.

7

Explaining ring phenomena

7.1 Saturn's broad rings

Saturn, with its rings, remains one of the truly beautiful objects in the sky. It is still one of the most requested and observed targets for small telescopes at observatories and planetaria. The ringed planet is also a popular symbol for the wonders of space. The magnificent images from Voyager and Cassini have assured that this will continue to be true for some time.

Some years ago, the broad structure of Saturn's rings was characterized by ground-based observers. Figure 7.1 is a tracing of ring brightness by Audouin Dollfus (1970) based on observations from 1948 to 1958. Almost every feature in this brightness profile has a clear basis in the optical-depth profile observed in the most recent studies; for comparison, Figure 7.2 shows the vertical optical depth from the Voyager 2 photopolarimeter stellar (PPS) occultation (Esposito *et al.*, 1983b).

One notable aspect from Figure 7.2 is the similarity of the inner edges of Saturn's B and A rings (at Saturnocentric distances of about 91 000 km and 121 000 km). These similar features have not been explained by any known resonance with Saturn's moons. One explanation is that the gradual ramps are formed from a balance between collisions among the ring particles and the production of debris from meteoroid collisions. The movement of ring material by impacts from meteoroids is termed *ballistic transport* (Durisen *et al.*, 1989, 1992, 1996; Cuzzi and Estrada, 1998). The fragments from a meteoroid collision have distinct orbits that re-intersect the rings. At a later ring crossing, the fragments collide with ring particles, imparting energy, momentum, and mass at that location. This transport modifies the expected ring evolution caused by slow spreading and resonant interactions, especially near the edges of opaque rings.

Two of the most recent views of Saturn's rings are shown in Figures 7.3 and 7.4: a Cassini image of sunlit Saturn and its rings (Figure 7.3) and the backlit

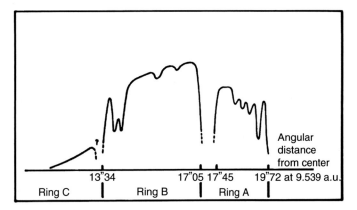

Figure 7.1 The overall structure of Saturn's rings as determined by ground-based observations. From telescopic observations at Pic du Midi in the years 1948–58. (Courtesy A. Dollfus.)

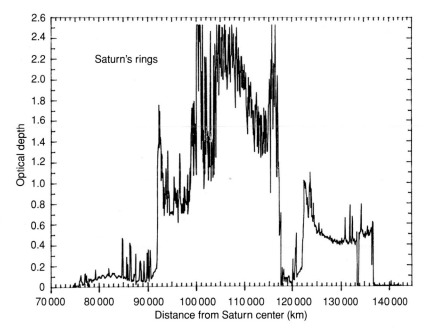

Figure 7.2 Optical depth of Saturn's rings measured by the Voyager Photopolarimeter Stellar (PPS) occultation. The resolution has been degraded by averaging the original data in bins of 600 measurements (about 50–60 km). (From Esposito *et al.* (1983b). Voyager PPS stellar occultation of Saturn's ring. *J. Geophys. Res.*, **88**, 8643–9.)

Figure 7.3 Saturn and its rings from the Cassini cameras. (PIA06193; courtesy of NASA/JPL/Space Science Institute.) For color version, see Plates section.

Figure 7.4 Saturn and rings in eclipse. Note the backlit rings, especially the diffuse E ring. (PIA08329; courtesy of NASA/JPL/Space Science Institute.) For color version, see Plates section.

planet and its rings from a vantage point of the planet's shadow (Figure 7.4). These images raise questions of the nature of Saturn's ring system: What causes the fine structure? What are the origin and history of these bright and massive rings?

Cassini confirms that Saturn's rings are dynamic and continually evolving. The rings change on timescales ranging from days to 10–100 million years, and our continuing experience confirms that each advance in observation reveals new structure. The ring particles are mainly aggregates of smaller particles arranged at many locations into transient elongated clumps tens of meters in size. Small moons near and within the rings are intimately involved in creating ring structure (see Chapter 6). Density waves excited by resonances with moons make up the majority of features in the A ring. Embedded moons create satellite wakes and perturb the edges of gaps cleared by these small moons. Propeller-shaped structures that show an intermediate stage in which the embedded object is not large enough to hold open a complete gap were theoretically predicted. They are now confirmed by Cassini images and occultations.

Understanding of the structure of Saturn's rings has improved dramatically over the course of the Cassini mission, in large part because of the combination of observations from multiple instruments taken from a wide range of geometries. This has enabled, for the first time, a detailed exploration of longitudinal structures in the rings as well as the vertical structure of the main rings. Not surprisingly, the macroscopic structure of Saturn's rings is nearly identical to that which was observed by Voyager 25 years earlier. The most notable exception is the change in morphology of the F ring, which has continued to evolve underneath Cassini's multiwavelength eyes. Saturn's D ring also shows significant changes.

The grand structure of the rings is shown in Figure 7.5, a composite of stellar occultation measurements and a Cassini imaging mosaic. Compare this with the Voyager results in Figure 7.2. Saturn's rings can be grouped into the classical dense rings (A, B, and C) and tenuous rings (D, E, and G). The Cassini Division separating the brightest rings A and B is not empty and resembles the C ring. Ring F shares some characteristics of both dense and diffuse rings, and contains a population of embedded objects (see Table 3.1).

The most common structures in the rings are density waves excited by the gravity of nearby moons, at locations where the ring particle motion is resonant with the moon, as explained in Section 6.4. Figure 7.6 shows the A ring and the outer Cassini Division. Each prominent wave is labeled with the ratio of the particle mean motion to that of the moon. For the Mimas 5 : 3 resonance, both the density wave and bending wave are evident (see Figure 6.1). Bending waves are a corrugation of the ring caused by resonance with an inclined moon.

In addition to density and bending waves, nearby moons can also create wakes and wavy edges in the rings. Figure 7.7 shows the variety of features in a region around Saturn's Encke Gap in the outer A ring. Pan, the moon whose gravity clears the gap (Showalter, 1991), is not visible in this particular image.

Density waves provide a local probe of the ring properties: the wave dispersion gives the surface mass density, and the attenuation of the wave amplitude gives the viscosity, which can be related to the interparticle collision velocity and the ring thickness (see equations 6.15 and 6.10; Tiscareno *et al.*, 2007; Colwell *et al.*,

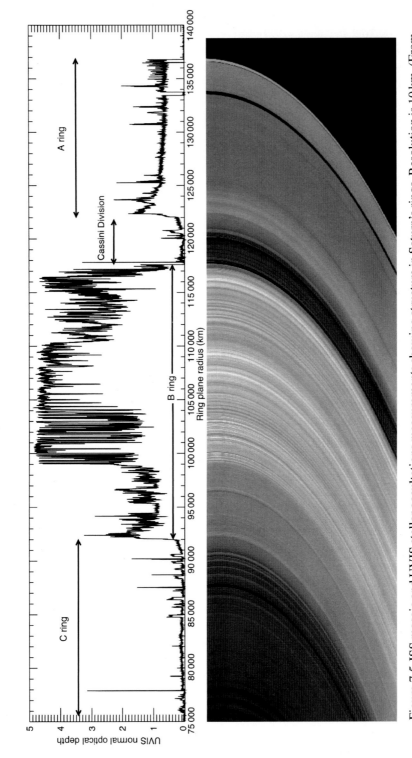

Figure 7.5 ISS mosaic and UVIS stellar occultations measurements showing structure in Saturn's rings. Resolution is 10 km. (From Figure 13.1, Colwell *et al.* (2009b); reprinted with kind permission from Springer Science and Business Media.) For color version, see Plates section.

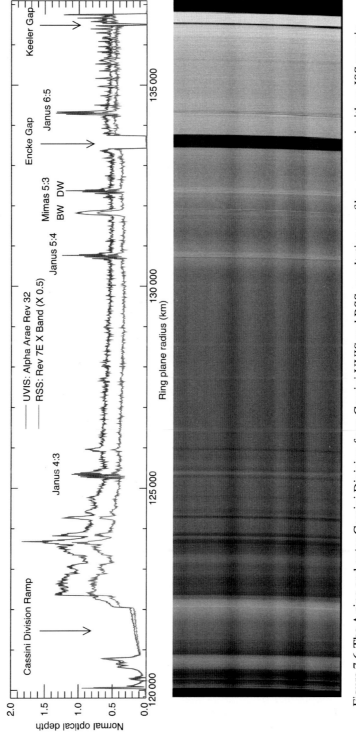

Figure 7.6 The A ring and outer Cassini Division from Cassini UVIS and RSS occultation profiles, compared with an ISS mosaic. Resolution is 6 km (image) and 10 km (occultations). BW, bending wave; DW, density wave. (From Figure 13.2, Colwell *et al.* (2009b); reprinted with kind permission from Springer Science and Business Media.) For color version, see Plates section.

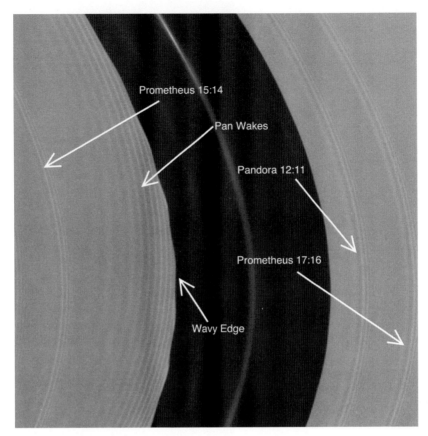

Figure 7.7 The Encke Gap (320-km width) imaged by Cassini at Saturn Orbit Insertion. Note the dusty ringlets within the ring, a wavy inner edge recently perturbed by the satellite Pan (which had previously passed through the image and is now well above this region) and satellite wakes on the inner edge. Density waves are labeled by the resonance that launches them. (From Figure 13.4, Colwell *et al.* (2009b); reprinted with kind permission from Springer Science and Business Media.)

2009a). Figure 7.8 shows that the waves in the outer part of Saturn's rings indicate a density of 1–60 g cm^{-2} and a velocity of 0.3–10 mm s^{-1}, implying a thickness of 3–5 m in the Cassini Division and 10–15 m in the inner A ring.

A major Cassini finding was the observation and characterization of self-gravity wakes (Colwell *et al.*, 2006, 2007; Hedman *et al.*, 2007b) from the comparison of multiple star occultations. The first Cassini occultations gave variable values for the transparency of the ring depending on the viewing geometry! This longitudinal variation can be explained by the fact that the ring is not uniform but is instead clumped into temporary aggregations. Because the aggregations are sheared by the outward decrease in Kepler orbital velocity, they have a preferred elongation tilted

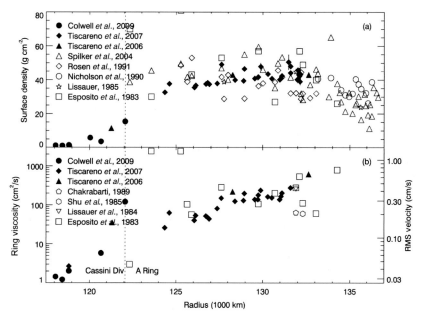

Figure 7.8 Surface density (*a*) and viscosity (*b*) of Saturn's A ring and Cassini Division, from analysis of density waves. Bottom right-hand scale is the inferred random velocity. All references noted in the figure can be found in Colwell *et al.* (2009b). (From Figure 13.3, Colwell *et al.* (2009b); reprinted with kind permission from Springer Science and Business Media.)

20–25° to the local orbital flow; this explains how the viewing angle yields different optical depth (Figure 7.9). One of several simple models compares the wakes to granola bars, by assuming they have a rectangular cross-section. No massive seed particle is required to form the wakes: unlike satellite wakes (see Figure 7.7), they are not wakes in the usual sense of the word (Colwell *et al.*, 2006).

The A-ring optical depth variations seen by the occultations are most pronounced in the central A ring, which is also the location of the peak amplitude of the *azimuthal brightness variation*, the long known phenomenon that the quadrants of Saturn's A ring are alternately a little brighter and a little darker than the average. This asymmetry (Camichel, 1958; Lumme *et al.*, 1983) provided the first clue to the existence of these self-gravity wakes. Simulations by Salo (1992, 1995) demonstrated that such wakes arise naturally in a self-gravitating ring. Porco *et al.* (2008) and French *et al.* (2007) demonstrate that the wakes can reproduce Voyager, Cassini, and Hubble observations of the A-ring brightness asymmetry. Their numerical results indicate that ring particle collisions must be dissipative, with $\varepsilon \ll 1$ (see Chapter 4).

Detailed models of the Cassini occultations (Colwell *et al.*, 2006) and thermal emission (Leyrat *et al.*, 2008) show the wakes are highly flattened, with a

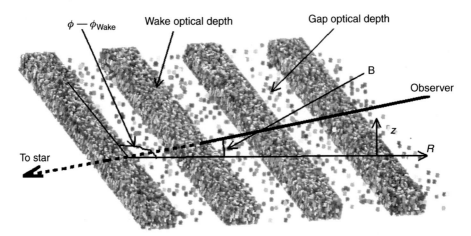

Figure 7.9 The so-called granola-bar self-gravity-wake model. Model parameters affecting measured optical depth are indicated. The blue line is the line of sight to occulted star. The wake orientation is given by $\phi - \phi_{\text{wake}}$. The star declination is B. For more details, see Colwell *et al.* (2006). (Reprinted with permission from *GRL*.) For color version, see Plates section.

height-to-width ratio of 0.2 to 0.4. Between the wakes are nearly transparent gaps with optical depth 0.1 to 0.3. These findings mean that previous interpretations have significantly underestimated the amount of material in the rings. Because of the self-gravity wakes, the average transmission is determined primarily by the width and transparency of the narrow gaps between the wakes (Colwell *et al.*, 2007), rather than the total cross-section of particles. Most of the light transmitted by Saturn's rings comes through the gaps. For practical purposes, the gaps are almost empty. Numerical simulations that incorporate a range of particle sizes (e.g., Salo and Karjalainen, 2003) predict that the larger particles are concentrated in the wakes, with smaller centimeter-sized particles distributed more evenly. This pre-diction leads to a picture of Saturn's B ring in which broad, flat wakes of densely packed meter-size particles lie in a thicker, low-density haze of small particles. The measured optical depth may therefore further underestimate the amount of ring material because it does not reflect particles hidden within the opaque B-ring wakes (Stewart *et al.*, 2007; Robbins *et al.*, 2010).

Figure 7.10 displays a simulation survey of wake structures expected at different planetocentric distances. The figure indicates clearly the gradual increase in the strength of wakes as the assumed distance or optical depth increases, as well as the increase in the clumpiness of the wakes, and their eventual collapse into aggregates at large distances.

We know that moons embedded in the rings can open gaps (see Section 5.2), as Pan creates the Encke Gap and Daphnis creates the Keeler Gap. It was expected

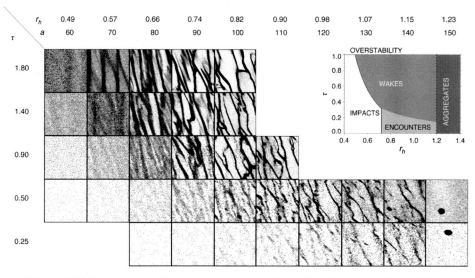

Figure 7.10 Simulations of ring structure for values of optical depth (τ) and distance from Saturn (a) in units of 10^3 km. The relative strength of self-gravity to tidal force is given by the parameter r_h (ratio of Hill radius to particle radii) for particle mass density $0.9\,\mathrm{g\,cm^{-3}}$. (Upper left) Axisymmetric overstable occultations coexist with inclined self-gravity wakes. (Inset) Summary of the locations at which processes dominate. (From Figure 14.7, Schmidt *et al.* (2009); reprinted with kind permission from Springer Science and Business Media.)

that small moons would explain the multiple gaps in the C ring and the Cassini Division. However, to date, no moons have been found in those regions.

Instead of finding these suspected moonlets in empty gaps, Cassini indirectly discovered evidence for moonlets that only partially open gaps. If an embedded moonlet is smaller than a few hundred meters, then the gap it induces in the rings may be closed by viscous diffusion before it extends completely around the ring. The competition between gravitational scattering and viscous diffusion creates a typical structure shaped like the letter S, which has been termed a propeller (Spahn and Sremcevic, 2000). Such S-shaped density undulations, induced in a disk as the local response to an embedded mass, were first studied by Julian and Toomre (1966).

Cassini images indicate millions of these propeller-shaped structures, showing incomplete gaps that can stretch for thousands of kilometers. These propeller moonlets may be evidence for limited accretion in the rings, a process that could recycle the ring material by storing it in these small moons, until they are later shattered to form new rings.

The moonlets are unseen, but their size is inferred from theoretical models of the propeller structure. They range in size from roughly 100 m to 1 km, roughly the

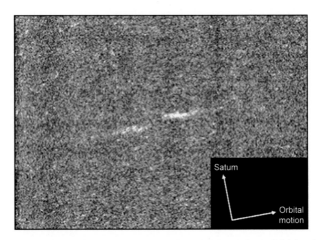

Figure 7.11 Propeller feature observed by Cassini on July 1, 2004. (Courtesy of NASA/JPL/Space Science Institute.)

same size as the *kittens* in Saturn's F ring (see Section 11.4) and structures seen in density waves and at the outer edge of Saturn's B ring in power spectral analysis of star occultations (Esposito *et al.*, 2012).

The first four propeller features (Figure 7.11) were seen in Cassini images by Tiscareno *et al.* (2006a). They were the first of a large number of propellers found in several belts in the middle part of the A ring (Sremcevic *et al.*, 2007; Tiscareno *et al.*, 2008). The exact interpretation of the propeller features is still disputed: Do the brighter regions seen in the images indicate more or less material? Photometric modeling by Sremcevic *et al.* (2007) interpreted the bright regions as wakes of enhanced density where the brightness is increased by release of regolith from colliding particles (Figure 7.12). Alternatively, Tiscareno *et al.* (2010) suggested that the bright regions are the result of disruption of self-gravity wakes downstream of the embedded moon.

These propeller-creating objects show a very steep size distribution, with power-law index of $q = 6$ (Tiscareno *et al.*, 2008). They therefore appear to be a distinct distribution from the majority of ring particles which have a shallower size distribution with $q \approx 3$ (equation 4.2). Furthermore, the moons Pan and Daphnis (which open the Encke and Keeler Gaps) do not fit this distribution. Schmidt *et al.* (2009) suggests three families of bodies in the rings: ring particles ($1 \, \text{cm} < a < 10 \, \text{m}$; see equation 4.2), propeller moonlets ($10 \, \text{m} < a < 500 \, \text{m}$), and ring-moons ($0.5 \, \text{km} < a < 100 \, \text{km}$), consistent with the scenario of ring formation in episodic cascades where ring-moons and moonlets are continuously destroyed by meteoroid impacts (Esposito *et al.*, 2005; Chapter 10). The steep moonlet size distribution may be essential to forming a visible propeller; otherwise, frequent

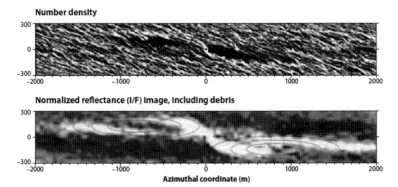

Figure 7.12 Numerical models of propellers. (Upper panel) Particle density distribution perturbed by 40-m-diameter moonlet. (Lower panel) Synthetic image for comparison with Figure 7.11, assuming loosely bound regolith, temporarily shaken loose by collisions. In this model, the released debris hides the predicted gaps in the numerical model and enhances the brightness of the density crests. (From Schmidt *et al.* (2009); reprinted with kind permission from Springer Science and Business Media.)

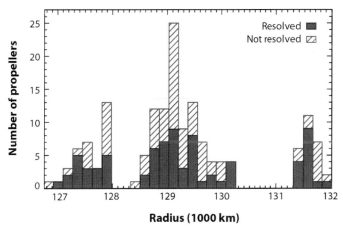

Figure 7.13 Radial distribution of 100-m-diameter propellers. Three distinct belts are seen. (Figure modified from Tiscareno *et al.* (2008); reproduced by permission of the AAS.)

interactions with neighboring larger moonlets would destroy the propeller feature (Lewis and Stewart, 2009).

The propellers are concentrated into three bands of width roughly 1000 km (Tiscareno *et al.*, 2008; see Figure 7.13) at the same distance from Saturn where the azimuthal brightness asymmetry is maximum and where the self-gravity wakes are strongest. Are the wakes created by fragments of an earlier destroyed object,

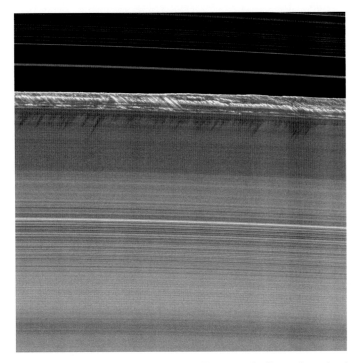

Figure 7.14 Vertical structures, among the tallest seen in Saturn's main rings, rise abruptly from the edge of Saturn's B ring to cast long shadows on the ring seen two weeks before the planet's August 2009 equinox. Part of the Cassini Division, between the B and the A rings, appears at the top of the image, showing ringlets in the inner division. Here, vertical structures tower as high as 2.5 km (1.6 miles) above the plane of the rings – this is much larger than the vertical thickness of the main A, B and C rings, which is generally only about 10 m. The bright reflections and long shadows may be created by large bodies up to a kilometer or more in size. This image and others like it are only possible around the time of Saturn's equinox, which occurs every half-Saturn-year, or about every 15 Earth years. The illumination geometry that accompanies equinox lowers the Sun's angle to the ring plane and causes structures jutting out of the plane to cast long shadows across the rings. (PIA11668; courtesy of NASA/JPL/Space Science Institute.)

or is this a region of increased accretion where aggregates have grown to sizes that are capable of creating propeller structures? This question is still under debate.

Exterior to the Encke Gap, only a few (generally larger) propellers are seen. The largest propellers clearly show the theoretically predicted incomplete gaps and the moonlet-induced wakes. Several models have been proposed to explain the non-Keplerian motion of some propellers (Rein and Papaloizou, 2010; Pan *et al.*, 2012; Tiscareno, 2013a). Future Cassini observations may be able to decide between the models. A similar effect (type III migration) has been suggested for

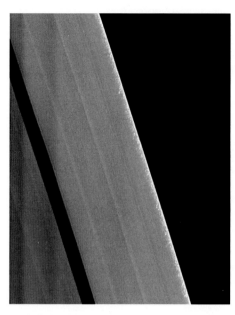

Figure 7.15 Clumps of ring material are revealed along the edge of Saturn's A ring in this image taken during the planet's August 2009 equinox. (PIA12722; courtesy of NASA/JPL/Space Science Institute.)

growing planetary embryos embedded in preplanetary gas-dust disks (Masset and Papaloizou, 2003; Papaloizou *et al.*, 2007).

At the equinox in August 2009, the Sun set on the rings. The oblique lighting highlighted any structures above the rings as they reflected sunlight, like the peak of a mountain at sunset. In addition, even objects too small to be seen cast long shadows that are big enough to resolve and show abundant structures at the edge of the B and A ring (Figures 7.14, 7.15). The small moons Pan and Daphnis also cast observable shadows (Figure 7.16) along with a number of previously unknown objects.

We still do not have a good explanation for the abundant *microstructure* in Saturn's rings (see Figure 7.17). The spacecraft data have demolished the earlier expectation that the rings were broad and homogeneous. Instead, we now know the rings are filled with numerous structures. Indications came first from the studies of the dark side of the rings and the ring-passage observations analyzed by Kari Lumme and William Irvine (1979), and from the Pioneer 11 data (Esposito *et al.*, 1980). These studies did not have the resolution of Voyager, but they showed light leaking through that could not have come through a homogeneous ring. Still, it was quite unexpected when Voyager approached Saturn for the first time in 1980

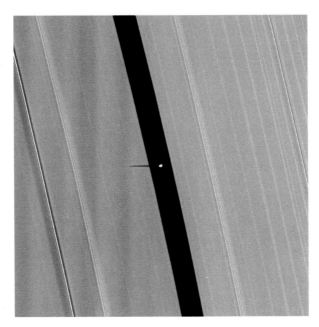

Figure 7.16 Saturn's small moon Pan, near the center of the image, casts a short shadow on the A ring in this image taken before the planet's August 2009 equinox. Pan (28 km across) orbits in the Encke Gap of the A ring. The equinox lowers the Sun's angle to the ring plane, significantly darkens the rings, and causes out-of-plane structures to look anomalously bright and cast shadows across the rings. These scenes are possible only during the few months before and after Saturn's equinox, which occurs only once in about 15 Earth years. (PIA11647; courtesy of NASA/JPL/Space Science Institute.)

(and even again for the second time in 1981 with the higher resolution provided by the stellar occultation) that we would continue to discover more structure in the rings. The structure is apparent at all ranges. The shape of the *power spectrum* is approximately 1/f down to the noise level: that is, we see structure at all scales in the rings, down to the best resolution we have (Voyager occultations: ~100 m: Esposito *et al.*, 1983a). Cassini UVIS star occultations now find meter-scale structures by observing stars whose apparent motion tracks the orbits of individual ring particles (Sremcevic *et al.*, 2013).

   A proposed explanation for this structure is *diffusional instability* (see, e.g., Lin and Bodenheimer, 1981; Ward, 1981); however, Jack Wisdom and Scott Tremaine (1988) have argued that this instability would not actually occur. Recently, Schmidt *et al.* (2001) suggested that this irregular structure may result from a viscous *over-stability* in the dense parts of Saturn's B ring. High-resolution studies from Cassini confirm this explanation for much of Saturn's A ring. The B-ring microstructure remains unexplained.

Figure 7.17 Small-scale structure in Saturn's B ring as seen by the Voyager camera at four different longitudes. Smallest features are seen about 20 km in width. NASA-JPL Voyager 2, 260–1473; see Smith *et al.* (1982). (Courtesy NASA/JPL-Caltech.)

7.2 Waves

Much of the structure in Saturn's A ring is explained by resonances with Saturn's small moons, which excite density waves in the ring. These resonances are mostly of the form $(m + 1):m$, the ratio between the moon's orbital period and that of the ring particle. The ring phenomena produced are tightly wound spiral density waves similar to those that cause the spiral structure of galaxies (Shu, 1984; see Chapter 6). The self-gravity of the rings provides a restoring force: this allows the wave to propagate away from the resonance, toward the perturbing moon. The Voyager PPS occultation and the radio occultation provided a very large set of waves – approximately 50 density waves and a small number of bending waves (Esposito *et al.*, 1983a, b; Esposito, 1986; Rosen, 1989; Rosen *et al.*, 1991a, b).

Planetary Rings

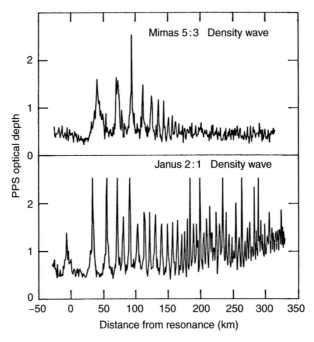

Figure 7.18 Two density waves in Saturn's rings. Note that the amplitudes are not small compared with the background optical depth and there is little resemblance to a pure sine wave. (From Esposito *et al.* (1983b). Voyager PPS stellar occultation of Saturn's ring. *J. Geophys. Res.*, 88, 8643–9.)

Figure 7.18 shows the Saturn Mimas 5 : 3 and Janus 2 : 1 density waves propagating outwards in the rings. Matt Tiscareno *et al.* (2007) has analyzed another few dozen waves seen in Cassini images. Why are these waves of interest? Simply, they provide a remote probe of the rings' nature. We cannot yet get our hands on the rings or even image an individual ring particle; thus, we are in the same situation as the *plasma* physicist in the laboratory: we cannot place our measuring instrument into the medium that we want to measure. These waves are diagnostics in the same way that waves sent through a plasma are plasma diagnostics – we analyze the dynamic response of the medium to these waves and we are able to infer a number of physical properties: mass density, viscosity, vertical thickness (see equations 6.15 and 6.16). As an example, the total mass of Saturn's rings can be estimated to be about the same as its small moon, Mimas (Esposito *et al.*, 1983a), assuming the ring particles have the same characteristics as those in the A ring, where most density waves are seen. For a number of reasons (opaque parts of ring B not penetrated by occultations, abundant clumping and gaps in the rings) we can now see this estimate as a lower limit to the ring mass.

The ability of the theory to reproduce the exact wave structure is more mixed. For the Mimas 5 : 3 density wave (Figure 7.18), the general shape matches theoretical models (e.g., Longaretti and Borderies, 1986). On the other hand, for the Janus 2 : 1 wave in Figure 7.18 (the longest wave train and also the first wave seen in the Voyager photopolarimeter data), we are still a long way from a complete explanation. Part of the wave behavior is understood. For example, the theoretical prediction for the wave crests' location is well matched by the waves that we see. The wave train includes 79 observed waves that are located as predicted by the linear theory (Shu, 1984). From this wavelength behavior, we estimate the ring mass density at the wave location to be about $70 \, \mathrm{g \, cm^{-2}}$ (Holberg *et al.*, 1982; Lane *et al.*, 1982; Esposito *et al.*, 1983a; see equation 6.16). The derived mass density of the rings does not seem to be changed very seriously by more detailed analyses, which include non-linear effects (see Shu *et al.*, 1985a, b). Tiscareno *et al.* (2006b) demonstrate how the 4–6 year orbital swap of Janus and Epimetheus forces a more complicated pattern, in agreement with the Cassini images.

7.3 Spokes

Thirty-seven days before its closest approach to Saturn, Voyager 1 discovered dark, nearly radial, wedge-shaped markings (spokes) in Saturn's B ring (Smith *et al.*, 1982; see Figures 7.19 and 7.20). It was quickly determined from Voyager high-phase-angle pictures that these spokes contain a large proportion of micron-sized particles.

The small size of the particles explains naturally why they appeared dark on Voyager's approach and bright after it: particles about the same size as the light wavelength will preferentially scatter light forward into the angles Voyager observed looking back toward the Sun (see Chapter 15). These small particles are also more easily moved by electric and magnetic forces, since their charge-to-mass ratio can be large. This consideration leads naturally to a possible connection with electromagnetic effects. The Voyager planetary radio astronomy experiment discovered both broadband electrostatic discharges (*Saturn electrostatic discharges*, *SED*) and a strong source of kilometric wavelength radiation (*Saturn kilometric radiation*, *SKR*). An interrelation between spokes, SED, and SKR seems plausible, since power spectral analysis of spoke activity (Porco, 1983) found a significant peak with period 640.6 ± 3.5 min, consistent with the 639.4-min period of rotation in Saturn's magnetic field. Thus, the spokes are preferentially created at a peculiar longitude of Saturn that is also known to emit SKR. The most plausible explanation for this correlation would be the existence of a magnetic anomaly in Saturn's *dipole* field creating the anomalous behavior of the *magnetosphere* near the rings.

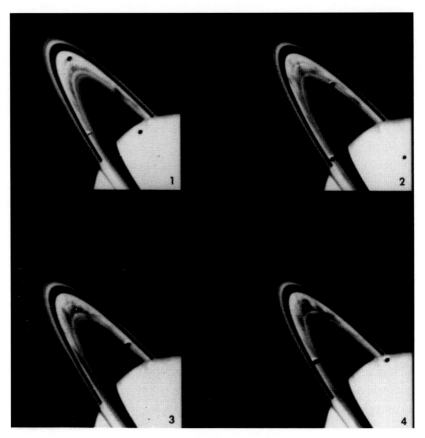

Figure 7.19 Voyager camera observations of spokes in Saturn's rings showing a range of spoke activity. Analysis shows that the strongest occurs at one partic- ular magnetic longitude. NASA-JPL Voyager 2 photos. (Courtesy NASA/JPL- Caltech.)

One of the most puzzling aspects of spokes is that they extend across the B ring for many thousands of kilometers. Spoke particles do not actually move radially but are observed close to their points of initial elevation. Bradford Smith *et al.* (1982) observed a narrow radial spoke grow to a length of 6000 km in less than 5 minutes. If the spokes are the manifestation of a discharge that proceeds along the length of a spoke, then this formation time implies a minimum disturbance speed of $2 \times 10^6 \, \mathrm{cm \, s^{-1}}$. This is much greater than any expected mechanical propagation speed for the rings; for example, velocity dispersion of the larger ring particles is less than $\sim 0.5 \, \mathrm{cm \, s^{-1}}$.

Voyager measurements shed some light on this. By measuring the motions of two narrow, forming spokes (including the one reported in Smith *et al.*, 1982), Eberhard Grün *et al.* (1984) determined that inside the corotation point, the trailing

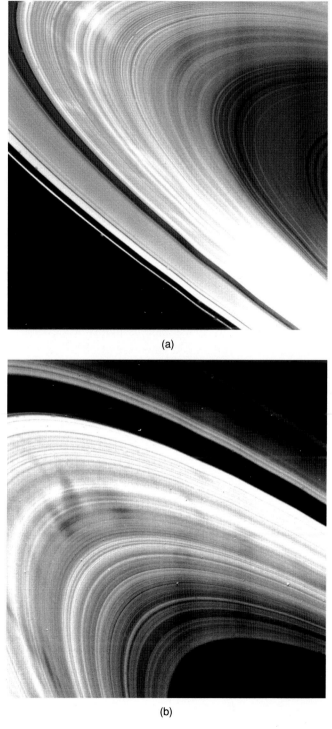

(a)

(b)

Figure 7.20 Spokes seen in forward-scattered light (i.e., backlit by the Sun) are now brighter than the surrounding ring; these similar Voyager images indicate that the spoke particles have a size near the wavelength of visible light. (a) NASA-JPL Voyager 1, PIA02269. (b) NASA-JPL Voyager 1, PIA02275. (Courtesy NASA/ JPL-Caltech.)

edges travel with the angular speed of the magnetic field, thereby remaining radial, while the leading edges move at Keplerian speeds. After formation, both edges tilt away from radial at the Keplerian rate. (Both edges of all other spokes measured in that study moved with Keplerian motion.) The angle of the wedge produced by the differential motion between edges is taken to be a measure of the time during which the spoke was active. Typical active times inferred from wedge-angle measurements of old spokes were 1 to 3 hours. Thus, spokes form in minutes and persist for hours. A possible cause is the impact of meteoroids on the rings, which creates a small puff of ionized gas (Morfill *et al.*, 1983a, b). One possibility recently proposed is that the spokes are the manifestation of magnetosonic waves in the partly ionized ring disk. This could explain their rapid propagation following the impact of a meteoroid (Tagger *et al.*, 1991). Jeffrey Cuzzi and Richard Durisen (1990) propose that the spokes are more likely to occur near dawn (as observed) because meteoroid impacts are more energetic on the night side of the planet where the ring-particle orbital velocity adds to Saturn's orbital velocity.

Cassini finally sees spokes. The many planned observations of spokes in Saturn's rings by Cassini in its first few years detected exactly none. This remarkable disappearance was confirmed by Hubble observations (McGhee *et al.*, 2005). Some initial models explained the disappearance as purely a viewing effect (McGhee *et al.*, 2005), but Cassini's orbit should have brought it to the appropriate viewing geometry in October 2004. Remarkably, no spokes were evident at that opportunity either. Colin Mitchell and Mihaly Horanyi (Mitchell *et al.*, 2006) concluded that spokes are indeed absent in the current epoch. They argue that the disappearance is due to a seasonal effect. When the Sun is high, as during the early Cassini years, a strong sheath of photo-electrons is produced above the rings. In this environment, any small grains levitated above the rings would be rapidly recaptured by larger particles they had escaped. Thus, no spokes would be visible. At lower Sun elevation, nearer to the time of the equinox on Saturn, the sheath weakens and the released grains can form the long, dark lanes seen by Voyager and Hubble. In line with this prediction, spokes reappeared in 2008. A recent study of the spokes seen by Cassini is by Mitchell *et al.* (2013). They confirm that Cassini sees spokes appear rapidly and evolve, and that most appear similar to those seen by Voyager, but do not decide between the two formation models. The Cassini cameras continue to examine the seasonal phenomena. See a more complete discussion in Chapter 16.

7.4 Narrow and eccentric rings

Until the discovery of the Uranus rings in 1977 (Elliot *et al.*, 1977; see Chapter 2), it seemed natural that all rings should be circular. Mutual collisions would cause eccentric motions to damp out and circularize the particle orbits. Unexpectedly,

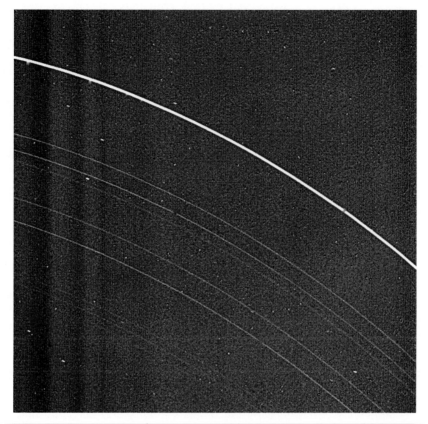

Figure 7.21 Uranian rings are seen in an image obtained by Voyager 2 from a distance of about 1 000 000 km as the spacecraft approached Uranus. The threadlike rings, which are for the most part densely packed with particles, are only a few kilometers wide. They are separated by hundreds of kilometers of virtually empty space. A new ring originally designated 1986U1R (now known as the λ ring) is barely visible between the outermost ring (ε) and the next bright ring (δ). The rings reflect only 1% of the incident light. (NASA-JPL Voyager 2, PIA01977; courtesy NASA/JPL-Caltech.)

the rings of Uranus (Figure 7.21) instead turned out to be eccentric, inclined, narrow, and sharp-edged. Most of these oddities can now be explained as due to the effects of nearby satellites (Goldreich and Tremaine, 1982; Borderies *et al.*, 1984; Dermott, 1984). To avoid the smearing out of an elliptical ring, Goldreich and Tremaine (1979) invoked the self-gravity of the ring particles themselves to create a precession of the ring that just counteracts the precession caused by the planet's gravity. This makes a very explicit prediction for the mass density of the rings of Uranus that is, however, not confirmed by the spacecraft observations (Goldreich and Porco, 1987; Esposito and Colwell, 1989; Esposito *et al.*, 1991;

French *et al.*, 1991). It is likely that particle collisions can play a critical role. A possible mechanism involving collisions is discussed by Eugene Chiang and Goldreich (2001).

The Uranus rings have provided a prototype for understanding other narrow, eccentric features in Saturn's rings (e.g., Porco *et al.*, 1984a, b; Porco and Nicholson, 1987; Porco, 1990). Likewise, close study of the Saturn features by Cassini has given some insights that will help explain some of the puzzles of the eccentric Uranus and Neptune rings. Cassini has discovered new rings created by small moons in the ring system. Because of the small sizes of the ring particles knocked off the source moons, they are subject to the multiple gravitational, radiation, and electromagnetic forces – still not completely deciphered. For more discussion, see Chapter 16.

7.5 Dusty rings

Dust is found in all the planetary ring systems (Burns *et al.*, 1984; Ockert *et al.*, 1987; Smith *et al.*, 1989; Esposito *et al.*, 1991; Showalter *et al.*, 1991; Burns *et al.*, 2001), generally in faint bands most easily visible when backlit by the Sun (indicating that we are seeing particles of about 1 μm or less in radius). The majority of the dust is concentrated near the equator. The dust in these rings is probably derived from nearby satellites (Burns *et al.*, 1980; Colwell and Esposito, 1990a, b; Burns *et al.*, 1999), and dust may also coat the moons' surfaces. The dust dynamics are determined by the planet's gravity and magnetic field, solar radiation pressure, local plasma, and gas density (Horanyi, 1998). If small enough dust particles become charged, they corotate with the planetary magnetic field. Radiation from the Sun can charge the dust grains and cause a relativistic drag force (Mignard, 1984). Furthermore, the dust is ground down by a continual flux of *micrometeoroids* (Burns *et al.*, 1980) and soon lost.

The premier "dusty" ring surrounds Jupiter (Figure 7.22), whose light is dominated by scattering from small particles. The dust is derived from small satellites (Ockert-Bell *et al.*, 1999), has its orbits modified by orbital evolution due to plasma drag, and spreads vertically and horizontally. Just as the gravity of moons sculpts the broader rings of macroscopic particles, resonances with Jupiter's magnetic field may cause boundaries and other features in the ring (Schaffer and Burns, 1987) and perturb the particles into inclined orbits, which explains the observed vertical thickness of the ring halo (Schaffer, 1989).

Galileo observations (Figures 3.2 and 7.23) have refined our picture of Jupiter's rings, showing clearly its three components: (1) a radially confined and vertically extended halo that rises abruptly; (2) a 6500-km-wide flattened and patchy main ring whose outer edge is bounded by the orbit of Adrastea; and (3) a pair of

Figure 7.22 One of the most spectacular of the Voyager 2 images obtained from inside the shadow of Jupiter. Looking back toward the planet and the rings with its wide-angle camera, Voyager took these photos on July 10, 1979, from a distance of 1 500 000 km. The ribbon-like nature of the rings is clearly shown. The planet is outlined by sunlight scattered from a haze layer high in the atmosphere. On each side, the arms of the ring curving back toward the spacecraft are cut off by the planet's shadow as they approach the brightly outlined disk. (NASA-JPL Voyager 2 photo mosaic; courtesy NASA/JPL-Caltech.)

exterior gossamer rings derived from the satellites Amalthea and Thebe, whose orbits circumscribe these rings (Burns *et al.*, 1999, 2001).

The phase curves of Jupiter's main ring (see Figure 7.24) are compatible with a power-law size distribution with index $q = 2.5 \pm 0.5$ (Showalter *et al.*, 1987; Brooks *et al.*, 2004). Galileo NIMS spectra at longer wavelengths (McMuldroch *et al.*, 1999; Brooks *et al.*, 2002, 2004) show that this power law steepens at sizes $r \geq 10\,\mu m$, perhaps indicating erosion of the particle size distribution by drag effects (see Chapter 14 for a more detailed discussion).

Dust is also prominent in Saturn's E, F, and G rings (Smith *et al.*, 1982), in Neptune's rings (Smith *et al.*, 1989), and between the rings of Uranus (see a comparison in Figure 16.1). However, dust is mostly absent from Saturn's main rings except in the localized spokes. Saturn's E ring is dominated by $1\,\mu m$ grains derived from Enceladus (Baum *et al.*, 1981; Showalter *et al.*, 1991; Hamilton and Burns, 1994). Some more detailed discussions can be found in Chapters 14 and 16.

7.6 Clumpy rings

Prior to the spacecraft observations of the past 30 years, it was possible to hold an idealized model of rings, which among other attributes assumed that rings

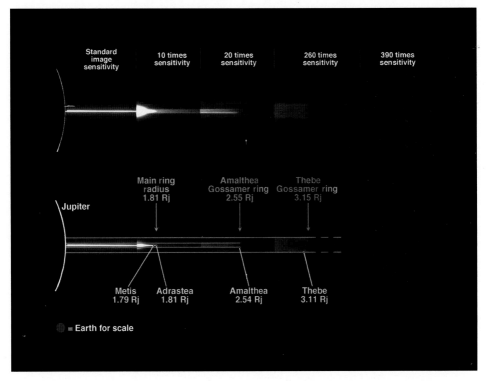

Figure 7.23 The Galileo mosaic C10, stretched to clarify individual elements of Jupiter's rings. The bottom frame is labeled. Both frames show clearly the faint halo, the gossamer rings, and their crudely rectangular end profiles. See Ockert-Bell *et al.* (1999). (PIA01623; courtesy NASA-JPL Galileo.) For color version, see Plates section.

possess azimuthal symmetry. Some azimuthal variations were known in Saturn's A ring (Camichel, 1958; Lumme *et al.*, 1977), but the actual extent of longitudinal heterogeneity was mostly unexpected. We have now observed plentiful examples. Among this group, which includes Saturn's F ring and the Uranus η ring, two cases are prominent: Saturn's Encke Gap ringlet (Figure 7.25, see also Figure 16.10) and Neptune's Adams ring (Figure 7.26). The irregular, kinky structure of the Encke Gap ringlet was attributed to small nearby moons; and in fact, a moon, Pan, was found there in a reanalysis of the Voyager camera images (Showalter, 1991). The Neptune ring structure includes a set of five small arcs in a longitudinal range of only 25° (Porco, 1991). The Voyager pictures show a very dim but continuous complete ring underlying these arcs, which was unfortunately too transparent to be detected from Earth-based occultations (Nicholson *et al.*, 1990). A model of Goldreich *et al.* (1986) explained the arcs through particles caught in corotation resonances with large, nearby inclined moons. The Voyager pictures did not show

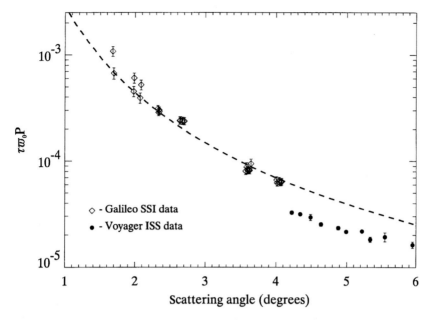

Figure 7.24 Jupiter's main ring phase curve. The ring brightness is plotted against the angle between the illumination direction and the viewing direction. (The scattering angle is 180° minus the phase angle.) The offset between the Voyager and Galileo data may be due to errors in the calibration of their relative sensitivity. The Galileo images extend the Voyager data to smaller scattering angles and show a similar power-law distribution for the Jupiter main ring particles. For reference, a theoretical phase curve for particle size distribution with $q = 2$ is fitted through the Galileo camera observations. (After Brooks *et al.*, 2004.)

any moons as large as predicted, but the radial excursions of the ring particles clearly show the influence of Neptune's moon Galatea, and the particles' azimuthal distribution was believed to be consistent with Goldreich's model (Porco, 1991). Unfortunately, collisions between the ring particles are likely to eject them from the weakly confining resonances, and we must consider this model incomplete. A possible improvement is that additional unseen moons can assist in this confinement (Sicardy and Lissauer, 1992), similar to the moon Pan's effect on the Encke ringlet (see Sections 13.3–13.7 for a more complete discussion).

The azimuthal structure in Saturn's narrow, slightly elliptical F ring (often referred to as *braids*) (Figure 7.27) can also be attributed to unseen moons (Kolvoord *et al.*, 1990; Kolvoord and Burns, 1992). In the model of ring creation by destruction of a moon, these hypothesized nearby moons may be the largest fragments of a satellite that was shattered to form the ring (see below).

The F ring was discovered by the author as part of the Pioneer 11 imaging photopolarimeter team (Gehrels *et al.*, 1980). The clumpy nature of the ring was

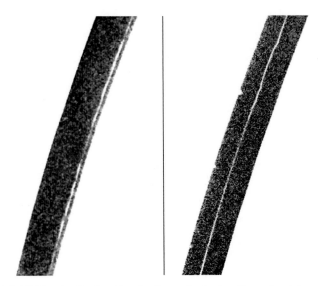

Figure 7.25 Within the Encke Gap in Saturn's A ring lie at least two discontinuous ringlets. This Voyager 2 picture, with a resolution of about 10 km, shows
one of these rings to be kinky. The kinks are spaced about 700 km apart, approximately 10 times more closely than the F-ring kinks photographed by Voyager 1.
(NASA-JPL, PIA01381, Voyager 2; courtesy NASA/JPL-Caltech.)

confirmed by Voyager (W. H. Smith *et al.*, 1981; B. A. Smith *et al.*, 1982). Although
it originally appeared to be a clear example of a shepherded ring (see Chapter 6)
since it lies between two nearby moons, Pandora and Prometheus, current opinion
is not so clear. Contradicting our expectations, the torques from these moons on
the ring do not actually balance (Showalter and Burns, 1982). Further, the orbits
of the shepherds are changing (Nicholson *et al.*, 1996; French *et al.*, 1999). One
suggestion is that their orbits are responding to as-yet-unseen moonlets within the
F-ring region, which may also be the source of the ring material (Cuzzi and Burns,
1988; Murray *et al.*, 1997; Barbara and Esposito, 2002; Murray *et al.*, 2008; Beurle
et al., 2010).

Clumps in the F ring appear abruptly, perhaps produced by impacts of interplanetary meteoroids into unseen parent bodies (Showalter, 1998; French *et al.*, 2012)
or from material knocked off by collisions of the parent bodies themselves (Barbara
and Esposito, 2002; Beurle *et al.*, 2010). Observations of Saturn's edge-on rings
using the Hubble Space Telescope in 1995 led to reports of previously undetected
moons outside the A ring. Unfortunately, no consistent orbits were found, and,
further, such moons should have been seen by Voyager. Most likely, they were just
more of those temporary clumps as seen by Voyager (Poulet *et al.*, 2000; Barbara
and Esposito, 2002). This conclusion provides good support for the model of Cuzzi

Figure 7.26 This Voyager 2 image shows the Neptune ring system in forward-scattering geometry (phase angle of 134°). Clearly seen are the three ring arcs and the Adams and LeVerrier rings. The direction of motion is clockwise; the longest arc is trailing. The resolution in this image is about 160 km. (NASA-JPL Voyager 2, PIA02207; courtesy NASA/JPL-Caltech.)

and Burns (1988), where a belt of smaller (0.1–10 km) moonlets fills a 2000-km band surrounding the F ring. Analysis by Amanda Bosh *et al.* (2002; see Figure 7.28) shows that the color of the F ring is compatible with the power-law size distribution with index $q \sim 2$. This value is similar to those from Jupiter's main ring (Brooks *et al.*, 2002, 2004) and Saturn's G ring (Throop and Esposito, 1998). French and Nicholson (2000) found that power-law size distributions fit the Saturn occultation data in the main rings with values in the range $2.7 < q < 3.1$ (see Table 7.1).

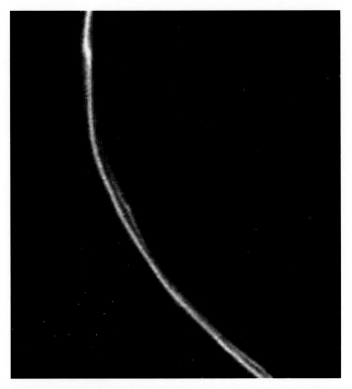

Figure 7.27 One of the most exciting photos of the Voyager 1 Saturn encounter
shows the F ring from a distance of 750 000 km. At a resolution of about 15 km, this
outer ring suddenly revealed a complex braided structure. Two narrow bright rings
that appear braided are visible, in addition to a broad diffuse component apparently
separated from them by about 100 km. Also visible is a kink, or knot, where the
ring seems to depart dramatically from a smooth arc. (NASA-JPL Voyager 1, FDS
34930.48, PIA02283; courtesy NASA/JPL-Caltech.)

Saturn's tenuous G ring, which is about 7000 km wide and centered on an
orbital radius of 168 000 km, is also derived from unseen parent bodies including
the recently discovered moon Aegaeon. Its light at all angles is dominated by dust,
however. Throop and Esposito (1998) favor a power-law size distribution with
$1.5 < q < 3.5$. Both Canup and Esposito (1997) and Throop and Esposito (1998)
demonstrate that this size distribution would be a natural result of disruption of a
progenitor parent satellite.

Saturn's E ring extends over hundreds of thousands of kilometers in radius, with
its brightness peaked near the orbit of Enceladus. Although it is believed that its
major source is from external impacts on that moon, we now know that it is created
from small ice grains erupted from volcanic geysers on Enceladus. They indicate

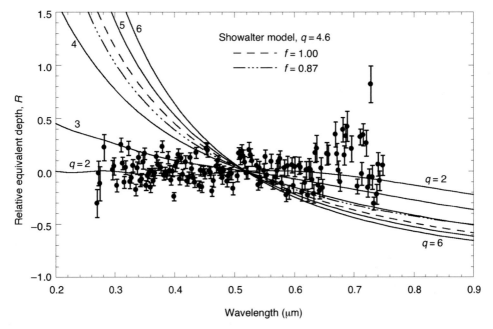

Figure 7.28 Relative total brightness of Saturn's F ring from Hubble Space Tele-scope data during the ring plane crossing of 1995. Models of Showalter *et al.* (1992) for dust fractions $f = 1.00$ and $f = 0.87$ are shown. Much better fits to the data are given by power-law size distributions with indices $q = 2$ or $q = 3$. (Figure from Bosh *et al.* (2002). Saturn's F ring. *Icarus*, **157**, 57–75, Elsevier (USA), reproduced by permission of the publisher.)

Table 7.1 *Regional particle size distributions*

Ring region	Radius range (km)	q	a_{min} (cm)	a_{max} (m)	a_{eff} (m)	$a_{eff}{}^{b}$ (m)	σ/τ_g (g cm^{-2})
C	74 490–91 983	3.1	1	10	2.3	≤1.2–2.8	149
B	91 983–117 516	2.75	30	20	8.3	5.7–8.8	816
Cassini Division[a]	117 516–122 053	2.75		20	7.0	≤1.1–4.5	582
		0.1					
A (inner)	122 053–133 423	2.75	30	20	8.3	11.2–12.2	816
A (outer)	133 745–136 774	2.9	1	20	6.0	9–10.7	455

Notes: q – power-law size index; a_{min}, a_{max} – smallest and largest particle in size distribution; a_{eff} – cross-section averaged mean particle size (Showalter and Nicholson, 1990); σ/τ_g – the ring surface mass density per unit geometrical depth, proportional to the particle density, assumed here to be $\rho = 1$ g cm^{-3}.

[a] The Cassini Division particle size distribution is not well determined from these obser-vations.

[b] Results from Showalter and Nicholson (1990) for the same ring.

Source: From French, R. G. and Nicholson, P. D., Saturn's rings II. Particle sizes inferred from stellar occultation data, *Icarus*, **145**, 502–23, ©2000 Elsevier Science (USA), repro-duced by permission of the publisher.

that the mass of all the visible material in the ring amounts to a moon merely 50 m across. See Chapter 16 for more details.

7.7 Summary

The structures and phenomena seen in planetary rings show a rich variety of physical processes, particularly in the interactions between small moons and nearby rings. This is confirmed by numerous images and stellar occultations from the Cassini space mission. In the following chapters, I discuss several approaches to explaining the ring phenomena, including their successes and failings.

8

N-body simulations

The many unusual ring features we have observed definitely call for some explanation. This requires detailed models. Perhaps the most straightforward way to make a model of the dynamics of a planetary ring is to simulate the individual motions and interactions of each of the particles. In principle, we can calculate the forces on each particle, solve Newton's laws to determine its trajectory, find its collisions with other particles, and calculate their outcomes. The simulation could be followed until a steady state or some other outcome is achieved. In practice, the memory and speed of computers limit this approach to considering only a small number of ring particles, say *N*. Thus, such calculations are referred to as *N-body* calculations. In the first such calculations, *N* might be only 100; nowadays 10 000 is common and 1 000 000 is occasionally achieved. This is still only a minute fraction of the number of ring particles in a planetary ring, so care must be taken to select characteristic samples or to scale the results to the real ring system.

Because this approach considers the evolution of all bodies, accurately numerically integrated, including gravitational forces and collisions, in principle no approximations are required. The ring dynamical processes may be described exactly. In practice, some approximations may still be necessary. Furthermore, following *N* particles is still a slow process for even the fastest computers: it is difficult to accumulate a broad range of solutions by testing many combinations of the parameters. Also, fluctuations of $\sqrt{N}$ may be expected. This may cause difficulty if we require high resolution or are interested in only some fraction of the ensemble of ring particles. For these reasons, in some cases statistical mechanics (see Chapter 4) or *stochastic processes* (Chapter 9) may be preferable.

N-body simulations are necessary for accurate treatment of many aspects of dense, flattened rings that include the collective self-gravity of the particles themselves. These conditions are beyond the scope for any analytical treatments.

Individual bodies and non-linear effects can be handled without resorting to the averages in the statistical approach.

The first simulations of planetary rings (Trulsen, 1971, 1972a, b; Brahic, 1975, 1977; Hameen-Anttila, 1978, 1981, 1982) considered a complete ring surrounding the planet. Each particle was on a Keplerian orbit. These early calculations confirmed the basic expectations of ring behavior. The ring flattens rapidly and then slowly spreads. Collisions between ring particles persist, even in this highly flattened state. The individual ring particles still have small random velocities (or equivalently small eccentricities and inclinations). An energy equilibrium is established where the energy lost in inelastic collisions is replaced from the shear of the particle motion as a result of Kepler's third law. The gradually spreading ring collectively supplies energy for the individual particles' velocity dispersion. This equilibrium depends on the detailed dependences of the inelasticity of collisions on the collision velocity, that is, particularly how the coefficient of restitution behaves with increasing velocity (see Chapter 4). These first calculations therefore provided important checks for analytic treatments and a confirmation of the basic physical understanding of dynamical processes in planetary rings.

8.1 Periodic boundary conditions

Modern N-body simulations (see, e.g., Salo, 1995; Lewis and Stewart, 2000, 2009) consider a small co-moving region within the rings. Because the ring particles closer to the planet are moving faster, any fixed region orbiting at the Kepler orbital velocity at its center will have particles leaving at some parts of the boundary and entering elsewhere. We can correct for this shear effect by "periodic boundary conditions," that is, a particle leaving at one boundary is replaced by another entering at the opposite boundary. This is equivalent to imagining that the simulation cell is surrounded by identical cells on all sides (see Figure 8.1). Since a much smaller region is now subject to detailed calculation, our same N bodies can go a lot further. The result is a much more realistic simulation, with much higher resolution, over a much smaller region of a planetary ring.

For each of the N particles, we solve the linearized equations of motion. For a box orbiting at a distance a from the planet, the Kepler velocity is Ωa. For x pointing away from the planet, y in the direction of orbital motion, and z upward, these are (Salo, 2001):

$$\ddot{x} - 2\Omega\dot{y} + (k^2 - 4\Omega^2)x = F_x \tag{8.1}$$

$$\ddot{y} + 2\Omega\dot{x} = F_y \tag{8.2}$$

$$\ddot{z} + \Omega^2 z = F_z \tag{8.3}$$

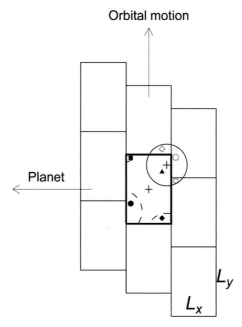

Figure 8.1 *N*-body simulation with periodic boundary conditions. Schematic diagram displaying the simulation cell (thick lines) and its eight surrounding replicas
(thin lines). Gravitational forces on a given target particle (denoted by the cross)
are calculated from all the other particles whose nearest image lies within the
distance R_{max} (denoted by the large circle). The nearest image can be either the
actual particle or one of its eight copies; an example of the former case is given by
the filled triangle, while the other filled symbols represent particles having one of
their images (open symbols) closer to the target particle. Broken curves indicate
regions from which particles contribute to the force felt by the target particle via
their images. L_x and L_y are the radial and tangential extents of the simulation cell.
(From Salo (1995); reprinted with permission from Elsevier.)

where the right-hand side represents the accelerations due to forces in addition
to the central planet's gravity, for example, the self-gravity of the ring particles
themselves.

Each particle crossing a boundary is replaced by an image particle at the opposite
boundary. Particles crossing the inner and outer boundaries of the co-moving box
have their y component of velocity adjusted for the Kepler shear. That is,

$$\Delta \dot{y} = \pm 3/2 \, \Omega L_x \tag{8.4}$$

where L_x is the x dimension of the box.

8.2 Impacts

The simulations generally assume instantaneous impacts between hard spheres. Linear and angular momentum is conserved, and the inelastic nature of the collision is handled by reducing the relative velocity after collision to a factor of ε times its value before the collision (ε is the coefficient of restitution; see Chapter 4). In some cases, slightly non-spherical particles are considered, as well as friction (which reduces the motion tangential to the particle surface to ε_T times its pre-collision value).

One of the largest problems for N-body simulations is to determine which ring particles actually collide. To do this rapidly is essential for the speed of the overall calculation, which determines the maximum number of particles N that may be considered in one simulation. The speed of the simulation depends critically on efficient search for collision pairs. One possible method is to find the very next impact, update the motions, and then repeat (this is the event-driven method). A faster method is to find all the collisions in the next simulation time-step. Heikki Salo (2001) uses a second-degree Taylor polynomial to identify the collisions, checking only those pairs that meet a threshold condition. One must be careful not to lose any particles or miss any collisions. Special care is necessary for the boundary crossers! Once the collisions are quickly identified, their actual trajectories are calculated very accurately with a better, higher-order integrator.

8.3 Results of N-body simulations

Salo (2001) summarizes a number of recent results. Simulations of a planetary ring composed of identical particles show that the velocity equilibrium is quickly established, independent of initial conditions. The final velocity depends on the coefficient of restitution. For the more realistic case of a distribution of particle sizes, N-body simulations show that the relative velocity and vertical extent of small particles exceeds the bigger ones, but without true "equipartition" of energy (see Figure 8.2). The smallest have 2% to 20% of the values predicted for complete equipartition.

The inclusion of self-gravity in the simulations has a significant effect. This shows that it is not possible to ignore the gravity of the ring particles relative to that of the central planet. Gravitational instability overshadows other effects. Realistic, highly flattened systems develop trailing wakes. The trailing behavior is a natural consequence of the differential rotation in a ring. By Kepler's laws, the ring particles further out rotate more slowly. Under some circumstances, this relative motion can stabilize the ring: particles are pulled apart faster than they would collect due to

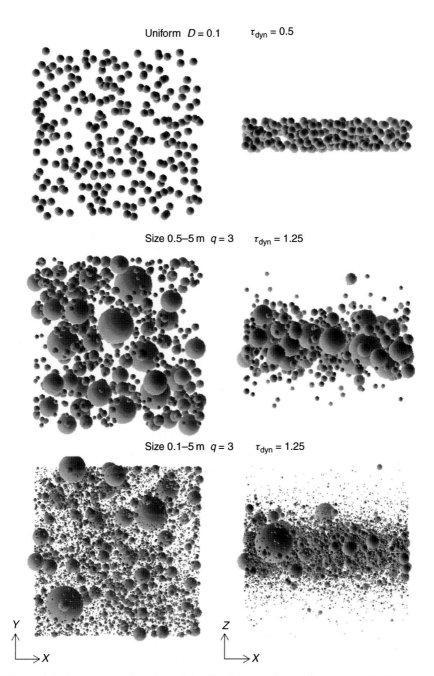

Uniform $D = 0.1$ $\tau_{dyn} = 0.5$

Size 0.5–5 m $q = 3$ $\tau_{dyn} = 1.25$

Size 0.1–5 m $q = 3$ $\tau_{dyn} = 1.25$

Y
X

Z
X

Figure 8.2 Some examples of particle distributions found from numerical simulations. In the upper row, a homogeneous system with volume density of 0.1 is shown, both as projected to the equatorial plane, and as seen from the side, along the direction of the orbital motion. The model was created by randomly placing the particles into the calculation region, with the condition that no overlaps are allowed. The two other examples are snapshots from dynamical simulations, using Bridges *et al.*'s (1984) elasticity law to describe the inelastic collisions. In all cases, just the innermost region of the particle field is displayed, having a factor of 2.5 larger actual extent in both planar directions. In the case with the most extended size distribution used (sizes ranging from 0.05 m to 5 m) the total number of particles is $\sim$200 000. (From Salo and Karjalainen (2003); reproduced with permission from Elsevier.)

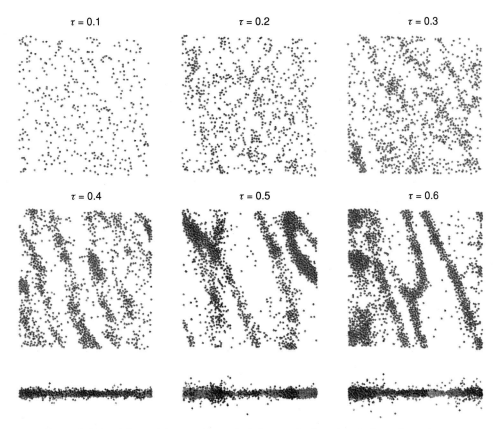

Figure 8.3 Particle positions at the end of the gravitating simulations. In each case $L_x = L_y = 112$ m and the different values of τ correspond to different N. The top rows show the view normal to the ring plane; the bottom row gives the side view. (From Salo (1995); reprinted with permission from Elsevier.)

their mutual gravitational attraction. The criterion for stability is due to Alar Toomre (1964), originally for galaxies:

$$Q_t = \frac{\kappa c}{\pi G \sigma} \geq 1 \qquad (8.5)$$

where κ is the epicyclic frequency, c is the mean dispersion velocity, G is the gravitational constant, and σ the surface mass density.

As can be seen, the ring system is stable if the density is low enough or the rotational velocity large enough.

Simulations show that these wakes (Figures 8.3 and 8.4) have a typical size of about 100 m across and an angle relative to the radial of about 25°. They last for a few periods (in Saturn's A ring, this amounts to a day or two) before dissipating. Wakes continually form and disappear. Models where the collisions are more elastic

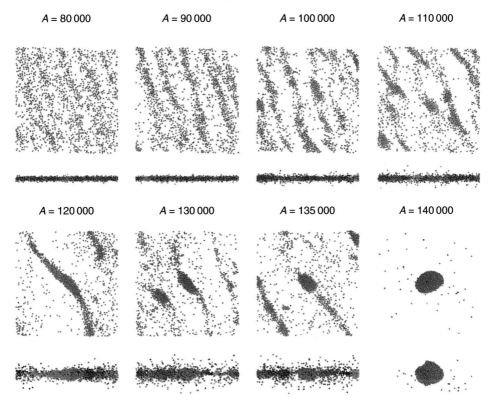

Figure 8.4 Particle positions at the end of simulations performed for various Saturnocentric distances (A in km) with $\tau = 0.4$, $\varepsilon = 0.5$, and $\rho = 900 \, \text{kg m}^{-3}$. The duration of each simulation was 15 orbital revolutions. In each case $N = 1600$, and $L_x = L_y = 112 \, \text{m}$. (From Salo (1995); reprinted with permission from Elsevier.)

($\varepsilon \approx 1$) lead to higher relative velocity for the ring particles (see Chapter 4) and suppress wake formation. The influence of a nearby moon can also perturb the ring, leading to larger particle speeds; this also suppresses the wakes (Lewis and Stewart, 2005).

It is likely that these wakes explain the fact that Saturn's A ring is fractionally brighter in some quadrants. This is the difference between looking across (bright) or along (darker) the wakes. Looking along the wakes, you see free space between them. Looking across them, the ring appears uniformly bright. These wakes are confirmed by Cassini occultations (see Colwell *et al.*, 2007, 2009b).

8.4 Stability, instability, and overstability

As discussed in Chapter 4, we can consider the random motions in a ring to be like a "temperature" in a gas. The basic source of this energy of motion in a

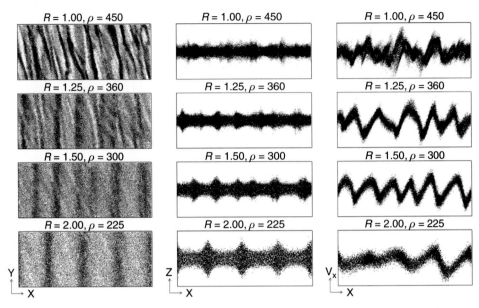

Figure 8.5 Snapshots from self-gravitating simulations after evolution of 50 orbital periods. In each case a local region of 583 m × 233 m is followed, corresponding to $10\lambda_{cr} \times 4\lambda_{cr}$. The simulation parameters are $\tau = 1.4$, $\sigma = 840\,\mathrm{kg\,m^{-2}}$, and the Saturnocentric distance is $100\,000\,\mathrm{km}$ ($\Omega = 1.945 \times 10^{-4}\,\mathrm{s^{-1}}$). The four different examples correspond to different combinations of particles' internal density ρ and radius R: the number of particles is between $15\,000$ and $60\,000$. In the leftmost column the system is shown from above (the planet is to the left, and the direction of the mean orbital motion is up), while in the middle column the system is shown from the side (the vertical extent of the frame is $\pm 0.25\,\lambda_{cr}$). In the right column the radial velocity profile is shown (vertical range is $\pm 0.8\,\mathrm{cm\,s^{-1}}$, corresponding to $20R\Omega$ for 1-m particles). The elasticity of impacts is described by the Bridges *et al.* (1984) formula. The self-gravity is calculated with an FFT, using an $n_x \times n_y \times n_z = 256 \times 64 \times 8$ density grid, combined with a pairwise calculation of particle–particle forces for mutual distances smaller than $\lambda_{cr}/5$. In each of the runs the Toomre parameter is $Q\tau \sim 1$, before the onset of overstability. (From Salo *et al.* (2001); reproduced with permission from Elsevier.)

ring is the Kepler shear in the particles' rotational velocity. As described above, this leads to a non-zero equilibrium temperature, as long as the coefficient of restitution decreases with velocity (more energetic collisions are less elastic). We can also consider the transfer of momentum between adjacent radial regions as a type of "viscosity." Terms like temperature and viscosity must be used with care, since the rings typically misbehave and are insufficiently described as fluids. Nonetheless, it is this viscosity that transforms the shared velocity into random motions or "temperature." If the viscosity decreases with the ring density, a possible instability can arise, called the *viscous instability.* The more dense regions balance against adjacent regions of low density. At the boundary, the momentum transport

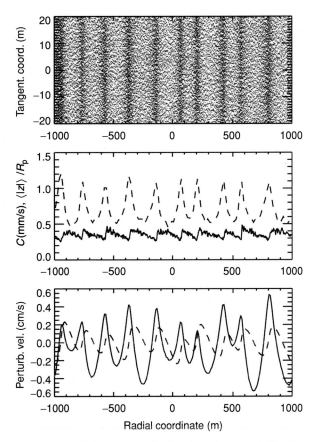

Figure 8.6 Perturbation of the state variables due to overstability in a simulation with optical depth $\tau = 1$ and particle radius 1 m after about 1000 orbital revolutions. The Saturnocentric distance was 100 000 km in this simulation and the self-gravity was mimicked by an enhancement of the frequency of vertical oscillations. The upper plot shows the particle positions projected in the plane $z = 0$ of the simulation box. Here, orbital motion is upward; the planet is to the left. Clearly visible are the radial perturbations on a 100-m length scale due to overstability. The plot in the middle shows the local velocity dispersion (solid line) and $\langle |z| \rangle / R_{\mathrm{p}}$ (dashed line), which is proportional to the local thickness, where R_{p} is the particle radius. The lower plot shows the perturbation in the radial velocity (solid) and in the shear corrected tangential velocity (dashed). (From Schmidt *et al.* (2001); reproduced with permission from Elsevier.)

is the same in both directions, despite the difference in density. In this instability, material flows into the dense region until the momentum transport balances. A similar instability may be found in the interstellar medium between the stars: thin, hot gas coexists with dense, cold regions. This mechanism was proposed after the early discovery of fine structure in Saturn's B ring (see Figure 7.17), called microstructure, by the Voyager spacecraft. However, it is now apparent that this particular instability does not occur: the coefficient of restitution leads to a viscosity that increases (not declines) with density. Furthermore, detailed *N*-body simulations show no evidence of this instability (Salo, 2001). Nonetheless, some possibilities remain. See Schmidt *et al.*, 2009.

Instead, the dense B ring of Saturn may be subject to the opposite effect, called *overstability* (Salo *et al.*, 2001; Schmidt *et al.*, 2001, 2009). In this case, flow is away from dense regions in the rings, but it overshoots. Small perturbations grow rapidly while maintaining their shape. The overstability would cause small random variations to grow rapidly into the chaotic structure that is seen (see Figures 8.5 and 8.6). *N*-body simulations show that the self-gravity of the ring particles is essential: no overstability occurs in models without self-gravity. This has now been directly observed in Cassini occultations.

8.5 Summary

Numerical simulations of the *N*-body type can, in principle, yield a complete picture of the structure and dynamics of planetary rings. They are not subject to the harsh approximations of analytic treatments and can treat all the phenomena exactly. Nonetheless, the long computing times are a limitation. These simulations clearly show that rings quickly attain an equilibrium velocity distribution or temperature, but are subject to instabilities due to the particles' own gravity. Gravitational wakes explain brightness asymmetry in Saturn's A ring. The overstability of the denser B ring may lead to the variety of structure seen there.

We ring researchers are now in a position to understand the collisional dynamics, origin of structure, and long-term history of rings. They form by fragmentation of a small moon, rapidly flatten into an equatorial disk, and spread slowly, sculpted by gravity from nearby moons. The ring particles are probably dynamic, ephemeral bodies like piles of rubble, which repeatedly aggregate and come apart. Collisions and meteoroid bombardment release dust, which is subject to more physical processes and thus is rapidly swept away.

Cassini observations show self-gravity wakes and overstability. Moonlets of size of 1 km or less are evident from A-ring propellers and temporary structures seen in Saturn's F ring.

Nonetheless, many details of origin and interaction, formation of radial and azimuthal structure, as well as individual particle characteristics are unknown. Discoveries in these areas could lead to major new understanding and concepts about planetary rings.

9

Stochastic models

In contrast to Chapter 8, we now consider models for planetary rings based on probabilistic evolution. I have discussed the detailed physical processes in previous chapters. It may seem odd, but a combination of both random and deterministic processes is essential to understand the history of planetary rings.

9.1 Motivation

Models of ring dynamics and history are most often deterministic. The dynamical model can be a numerical simulation that explicitly solves a differential equation. Alternatively, the simulation keeps track of each particle (this is the N-body simulation, where a set of N particles provides a sample of the ring evolution; see Chapter 8). Stewart *et al.* (1984) review kinetic approaches. Recent examples of N-body simulations are by Salo (1995) and Lewis and Stewart (2009). These models, although often idealized and simulating a limited time interval, clearly explicate the ring evolution. Nonetheless, it is clear that random events have played a significant role in the history of planetary rings. This goes beyond the normal fluctuations about the mean that are characteristic of other truly random processes such as diffusion in a gas (see, e.g., Landau and Lifshitz, 1969, Chapter 10). The importance of stochastic processes for rings is not just the random nature of the significant evolutionary events but the fact that the present nature of planetary rings is dominated by just a small number of recent events.

Contingency on a few random events may explain why each planet's ring system appears different. In each case, a small number of larger bodies may dominate the mass, history, and dynamics of ring systems. The rings themselves are likely to be the progeny of singular events like the destruction of a former moon or of a comet that passed too close. In the continuing competition between fragmentation and accretion, occasional major events set the pace. Transient features like arcs, clumps, and spokes are common. Altogether, ring history seems disorderly: rings display little of the Newtonian clockwork and clarity we often associate with the heavens.

These aspects of planetary rings mean that often we must account for random events and their ramifications in ring history. An approach continuous in time and space may overlook the jagged nature of the phenomenon. Furthermore, the present state of the rings is contingent on these past events: the details of each particular history cannot be ignored. The rings' current state may be just a particular sample that nature has drawn at random from a large range of similar (or perhaps dissimilar) possible outcomes. Thus, when we explain the ring observations, we must not only match what we see but also consider the likelihood of the particular outcome we observe.

My approach to handle these aspects is to consider ring dynamics and history as stochastic processes, that is, physical processes that are subject to random events and that evolve probabilistically in time. I briefly describe some useful concepts below. These concepts are used in a number of examples in following chapters.

9.2 Stochastic processes

The general idea of a stochastic process is a (discrete or continuous) sequence of random variables X_i (t_n) (in the continuous case, we have instead X_i (t)). These X_i are the numerical outcomes of the random events in the system and thus represent the values of some physical parameters of interest at times t_n (continuous: t). The random variables X_i have a probability distribution at time t_n that may depend on the past history of the system. A *stochastic process* is the mathematical abstraction of an empirical process whose development is governed by probabilistic laws.

9.3 Random walk on the line

As an example, consider a particle on the origin of the Cartesian x-axis. At constant intervals, it moves unit distance, either right or left, with probability p or q respectively ($p + q = 1$). This stochastic process, called the random walk (and sometimes the *drunkard's walk*), is an analog for the physical processes of Brownian motion and diffusion. If $p = q = 1/2$, the walk is symmetric; for $p > q$, the process shows a drift to the right. In such a drunkard's walk, each step is independent, that is, p and q do not depend on time or the drunk's current position.

9.4 Markov processes

The idea of a Markov process generalizes this idea of independence. For a *Markov process*, the conditional probabilities at a time t_n depend only on the state of the system at the time t_n and not on its previous history. The Markov process is a particular type of stochastic process that lends itself readily to linear algebra and matrices (see Kemeny and Snell, 1960; Esposito and House, 1978). The great value of Markov processes is that they are not only simply handled mathematically but they are also applicable to many practical problems, including the evolution of planetary rings. A finite Markov chain has a finite number of values of the random variable corresponding to discrete states of a physical system. At discrete time intervals, the physical system transitions to its next state (i.e., it attains a new value of the random variable at the next time-step). The probabilities for this transition depend only on the current state (this independence is the *Markov property*) and are independent of time. In this case, the state of the system is described by a *state vector* at any time (including, for example, the initial state, representing the initial conditions); and the transition probabilities can be arranged as a matrix. Successive multiplication of the state vector by the *transition matrix* gives the probabilistic evolution of the system with time. The state vector gives the expectation value for all the possible (finite) states of the system as the system evolves. This approach can be used even where a deterministic approach would also suffice. In Canup and Esposito (1995, 1997), we modeled the accretional growth of small bodies in the Roche zone with a Markov process that is entirely equivalent to solving the integro-differential equation for accretion. Brophy and Esposito (1989) modeled the collisional dynamics of narrow rings with a finite Markov chain; since this could also be cast as an *N*-body calculation or the solution of the Boltzmann equations for the kinetic ring system (see Stewart *et al.*, 1984), the benefits were purely practical. It was convenient for us to ignore the details of individual orbits and collisions and take a statistical approach using this formulation. If the motion of ring particles is chaotic, as for example in Neptune's rings (Foryta and Sicardy, 1996), this is a natural approach (Esposito *et al.*, 1997).

9.5 Markov chains and Monte Carlo simulations

Another way to simulate the evolution of a probabilistic system is to draw random numbers that determine the successive states of the stochastic process at each time-step. This is called the *Monte Carlo method*. If the random numbers are drawn against the conditional probabilities for a Markov chain, the state vectors calculated by successive multiplication give the expectation values for the Monte Carlo simulations after each step: thus the two methods are equivalent for the discrete-time, *discretized* (finite number of states) system. The Monte Carlo method gives a particular actualization of one possible history of the stochastic process. My students and I refer to this as "watching the video" of the Markov chain. If we are concerned with singular events, or outcomes with small-numbers statistics, then the

Monte Carlo simulation is what should be directly compared to our observations of planetary rings (see Colwell and Esposito, 1992). We have applied this method to estimate the growth of regolith and to meteoritic pollution of the rings (Elliott and Esposito, 2011).

9.6 Stochastic processes as ring models

Several good reasons justify using stochastic models. First, the actual physical processes are stochastic. In planetary rings, collisions, resonance trapping and escape, charging of dust by individual electrons, accretion, and fragmentation are all random events. Second, analytic solutions often do not exist and the partial differential equations may be difficult to solve. In this case, the stochastic models provide a numerically robust method for tracking the evolution of the system. Third, stochastic models simulate the actual time history (i.e., the "video") of systems that may be far from equilibrium and subject to singular events. Fourth, the probabilistic methods have a simple physical interpretation as the evolution of the state distribution of the key physical variables. It is easy to understand the results. Fifth, this approach emphasizes the contingent nature of the results. Each particular outcome is associated with its probability of occurrence. Every history may be different. In particular cases, we find that the dynamics are dominated by a limited number of the most massive bodies in the ensemble.

9.7 Summary

The stochastic approach can accommodate aspects of evolution that are difficult for a classical "clockwork" approach, for example, chaotic dynamics and catastrophic events. This is true when large fluctuations or jumps occur, for example, birth and death processes for small numbers. Gerd Röpke (1987) remarks that the deterministic transport equation like the *Liouville equation* is just a "degenerate" Markov process: the outcome is completely determined. Where small numbers dominate, as in the *collisional cascade* to form rings (Colwell and Esposito, 1992, 1993) or the sudden brightenings of Saturn's F ring (Barbara and Esposito, 2002; French *et al.*, 2012), the stochastic approach can capture these aspects. I will discuss these examples in more detail in the next chapter. This approach emphasizes the contingency of the present state: it very much depends on a few random events. In the case of ring evolution, these are, for example: the number of satellites, the distance to the Roche limit, and the characteristics of the most recent catastrophic destruction.

   A small number of larger aggregates, large fragments remaining from the disruption of an embedded moon, or periodic forcing from satellite resonances can agitate the ring system. Even very rare outcomes can be important, if these can become established, for example, if compaction creates more competent objects that are resistant to erosion and can only be destroyed by external impacts (see Esposito *et al.*, 2012). This contingency provides a philosophical connection to our current understanding of the demise of the dinosaurs or the formation of our Moon. Our present Moon and our present human dominance are contingent on rare, chance events like those that create planetary rings. In all these cases, the actual histories bear little relation to the ensemble averages. Our stochastic models (beyond their ease and practical advantages) can capture essential aspects of our solar system.

10

Age and evolution of rings

10.1 Introduction

As described by Harris (1984), the rings of the planets probably result from the same process that created the regular satellites. Like the ring particles, the satellite orbits are *prograde*, equatorial, and nearly circular. A question that immediately arises is whether rings are (1) the uncoagulated remnants of satellites that failed to form, or (2) the result of a disruption of a pre-existing object. A related question highlighted by the apparent youth of the rings is whether this latter process of ring creation by satellite destruction continues to the present time. This possibility thus mixes the origin of the rings with their subsequent evolution. Whatever their origin, the sculpted nature of the rings of Saturn, Uranus, and Neptune requires active processes to maintain them (see Chapter 13).

Because of the short timescale for viscous spreading, gas drag, particle *coagulation*, and transport of momentum to the forming planet, Alan Harris (1984) argues that rings did not form contemporaneously with their primary planets but were created later by disruption of satellites whose large size had made them less subject to the early destructive processes. The pieces of the disrupted satellite are within the Roche zone, where tidal forces keep them from merely coagulating. This explains naturally the presence of shepherd satellites and ring-moons around the various giant planets as the largest pieces remaining after the destruction.

Conversely, both Lissauer *et al.* (1988) and Wing Ip (1988) have shown that it is very unlikely for a moon large enough to create the Saturn rings to be disrupted by the meteoroid flux recorded on the surfaces of the remaining Saturn satellites. Harris (1984) notes that the ring disruption hypothesis is particularly attractive for the Uranus rings; below, arguments are presented that processes currently active in the Uranus and Neptune systems may require some very recent ring origins. For a recent review of the origin of Saturn's rings, see Charnoz *et al.* (2009).

10.2 Age of the rings

Estimates of the age of the rings can discriminate between possible scenarios for ring formation. If lifetimes of some components are much less than the age of the solar system, those parts cannot have a primordial origin. This argument indicates the recent origin of the material we observe in the Jovian ring (Burns *et al.*, 1980, 1999), Saturn's A ring (Esposito, 1986), Saturn's F ring (Cuzzi and Burns, 1988), the Uranus rings (Esposito and Colwell, 1989; Colwell and Esposito, 1990a), and the Neptune rings (Smith *et al.*, 1989; Colwell and Esposito, 1990b, 1992, 1993). Micrometeoroid bombardment of Saturn's rings is "interplanetary pollution," which should darken them rapidly (Cuzzi and Estrada, 1998).

The narrowness of the observed rings of Uranus and Neptune raised the first concern about the age of these rings. Because the interparticle collision time is short (less than half the orbital period for optically thick rings (Stewart *et al.*, 1984)), the particles collide and interchange momentum, causing the ring to spread (see Chapter 5). Modeling a narrow ring as a fluid of single-size particles, one can define a local kinematic viscosity (compare equations 4.8 and 4.9):

$$\nu \approx \frac{c^2}{2\Omega} \left(\frac{\tau}{1 + \tau^2} \right) \tag{10.1}$$

where c is the interparticle random velocity, Ω the mean motion, and τ the optical depth (Cook and Franklin, 1964). Even if the collisional velocity is minimized, the ring particles have a finite size, so that just the Kepler shear across a particle diameter will provide a non-local contribution to the viscosity for dense rings. For a more complete discussion, see Salo *et al.* (2001). The minimum viscosity would be achieved if random motions vanish, in which case the ring behaves as an incompressible fluid (Borderies *et al.*, 1985). This gives:

$$\nu_{min} = \Omega \left(\frac{\sigma}{\rho} \right)^2 \tag{10.2}$$

where σ is the surface mass density of the ring and ρ the density of the ring particle. Modeling the ring spreading as a diffusion process leads immediately to the lifetime of a narrow ring of width r. The lifetime is (compare equation 4.9):

$$\Delta t \approx \frac{\Delta r^2}{\nu} \tag{10.3}$$

For the Uranus rings we find $\sigma_{min} = 80 \, \text{g cm}^{-2}$ (Gresh *et al.*, 1989) for $\rho = 1.4 \, \text{g cm}^{-3}$ (average for the Uranus satellites: Tyler *et al.*, 1986), and $\Delta r = 100 \, \text{km}$ (maximum width of the ε ring) yielding $\nu_{min} > 1 \, \text{cm}^2 \, \text{s}^{-1}$ and thus $\Delta t \lesssim 10^6$ years. This calculation shows that these rings must be confined, most likely by shepherding satellites (Goldreich and Tremaine, 1980; see Chapter 5).

A major success of the Voyager encounter was the discovery of the satellites shepherding the ε ring (Smith *et al.*, 1986; Porco and Goldreich, 1987; see Figure 3.6). However, the apparent lack of other satellites to shepherd the other Uranus rings is also significant. This may be because the other shepherds were too small to be seen by the Voyager cameras ($R < 10$ km: Smith *et al.*, 1986). The shepherds (seen and unseen) could then confine all the rings, preventing the very short lifetimes expected from unconstrained spreading. More severe is the fact that Cassini has still not found shepherds for the many ringlets in the Cassini Division, indicating the incompleteness of this explanation (Hedman *et al.*, 2010).

10.3 Age of the Jupiter rings

The brightness of Jupiter's rings in forward-scattered light shows that many of the particles are about the same size as the wavelength of visible light. Such small particles can only persist for 10^3 years or less before being destroyed by *sputtering* or being dragged into the planet by interactions with the local plasma. The best way to resupply these rings is by meteoroid bombardment of small moons of Jupiter (Burns *et al.*, 1999) that eject dust into orbit around the planet. See more discussion in Chapter 14.

10.4 Age of the Saturn rings

Spreading of Saturn's A ring due to mutual collisions among the particles (Esposito, 1986) and darkening of the rings due to meteoroid material that coats their surfaces (Cuzzi and Estrada, 1998) both give ages much shorter than the solar system. It is hard to reconcile such youthful aspects with the large mass of Saturn's rings (equal to or larger than that of the moon Mimas). If Saturn's rings were created by the destruction of one of Saturn's moons, this would be a very rare event in solar system history and thus unlikely to have occurred in just the last billion years or less. An alternative explanation of the destruction of a close-passing comet (Dones, 1991) is also a very rare and unlikely event. Canup (2010) has suggested a Titan-sized differentiated moon was dragged inside Saturn's Roche limit, where the outer, icy mantle was stripped off, leaving the rocky interior to be eventually dragged into Saturn. The likelihood of such an occurrence has not yet been estimated.

10.5 Age of the Uranus rings

A strong argument for the youth of rings is related to the rates of momentum transfer between the objects in the Uranus ring system (Esposito and Colwell, 1989). These rates were first considered in light of the Voyager data by Goldreich and Porco

(1987). As discussed in Section 4.1, angular momentum flows outward through an unperturbed disk carrying the viscous torque (for particles in Keplerian orbits):

$$T_v = 3\pi \sigma v \Omega r^2 \tag{10.4}$$

The best explanation for the narrow rings of Uranus is that they do not spread because an inner shepherd supplies this torque at the inner boundary, while an external shepherd carries it off at its outer edge. For the Uranus ε ring, we know the sizes and locations of the two shepherds and can estimate their masses from their sizes (Porco and Goldreich, 1987). Lower limits to the mass of the ε ring can be inferred from the radio occultation results (Gresh *et al.*, 1989; Gresh, 1990). For σ greater than this σ_{min}, which is $80\,\mathrm{g\,cm^{-2}}$, the mass of the ε ring M_ε must be greater than $10^{19}\,\mathrm{g}$. Since this is only a lower limit, it leaves open the possibility that the ε ring is more massive than the shepherds! If the ring is more massive than the shepherds, transferring momentum to them cannot significantly slow the spreading. However, the mass of the inner shepherd Cordelia is some three times the minimum mass of the ε ring (with an uncertainty of perhaps a factor of 2: Porco and Goldreich, 1987), thus, as long as M_ε is close to its minimum value, this difficulty is avoided. Goldreich and Porco (1987) show that, in this case, the satellites are massive enough to supply and carry off the viscous torque T_v.

However, a lower limit to the viscous torque may be established by considering the drag on ring particles from the extended Uranus exosphere. The viscous torque must exceed the drag torque; otherwise the outer edge would be dragged inwards by the transfer of momentum as the ring particles plow into the exosphere. The excess torque, $T_{excess} = T_v - T_D$, is carried off by the resonance coupling to Ophelia. This momentum transfer causes Ophelia to evolve outwards away from the ring, and the ring spreads, maintaining its sharp outer edge at the slowly changing location of the Ophelia resonance.

The minimum viscous torque is thus $T_v = T_D$. At the ε-ring inner edge, a resonance transfers the momentum of the inwardly diffusing ring particles to Cordelia. This momentum transfer, $T_v + T_D > 2T_D$, causes Cordelia's semi-major axis to decrease. Since the mass of Cordelia is known, we can estimate how long would be required for Cordelia to have evolved inwards to its current separation, assuming the minimum torque and zero initial separation. The maximum duration of this shepherding is thus $t_{max} = \Delta L / (2T_D)$, where ΔL is the total change in angular momentum, and T_D is $9 \times 10^{16}\,\mathrm{ergs}$ (Goldreich and Porco, 1987). This calculation gives $t_{max} = 6 \times 10^8$ years, considerably shorter than the age of the solar system (Esposito and Colwell, 1989). Cordelia could not have transferred this momentum to a larger inner moon, since none exists. Further, it could not have transferred it to one of the larger outer moons through a resonance that may have existed in the past, since capture into a resonance with an outer moon is not possible (Peale, 1986), given the tidal expansion of the orbit of the larger outer satellite and the shrinking of Cordelia's orbit. This age, t_{max}, provides a strong upper limit because only minimum values for the torque, T_v, have been considered. In the minimum case where $T_v = T_D$, no torque is transferred to Ophelia while a torque $T_v + T_D = 2T_D$ is transferred to Cordelia, causing it to evolve inwards. Even shorter ages are implied for possible shepherds of the α and β rings.

One proposed solution to the short lifetimes of the Uranus rings is that dust, rings, and small moons are continually created by disruption of larger objects (Esposito, 1986; Esposito and Colwell, 1989; Colwell and Esposito, 1990a). Esposito and Colwell propose two families of objects too small to be seen by Voyager. Small satellites $R \lesssim 10\,$km are *ring precursors*. Belts of particles $R \sim 100\,$m are called *moonlet belts*. These are similar to the material proposed to create Saturn's F ring by Cuzzi and Burns (1988); see also Barbara and Esposito (2002), Esposito *et al.* (2012), Attree *et al.* (2012). Evolutionary calculations that include both fragmentation and accretion can yield bi-modal size distributions consistent with this expectation of larger bodies embedded in the rings (although this may require other processes to boost accretion; see Meinke *et al.*, 2012). Meteoroid ejecta from rings, moons, and moonlet belts are transported inward by gas drag, leading to a continuous low-optical-depth sheet of dust in the main ring system, with the highest optical depth near the ε ring, where gas drag is slower. At the location of a moonlet belt, a dust band would be visible. Esposito and Colwell (1989) propose that these processes, along with variations in the widths, optical depths, and size distributions of the moonlet belts, account for the variety of forms of dust bands seen by Voyager at Uranus, thus giving an explanation for the structure seen in Figure 16.1b.

10.6 Age of the Neptune rings

The azimuthal structure of Neptune's rings is another strong indication of youth. Although Goldreich *et al.* (1986) have shown how the arcs might be maintained, and Porco (1991) has shown that the Voyager observations are consistent with this model (but see Chapter 13), nonetheless, interparticle collisions will allow particles or their fragments to escape the resonance. Particles in orbits outside the small range of the corotation resonance will rapidly circulate to fill the entire ring circumference, creating a complete ring. Furthermore, recent Earth-based observations (Dumas *et al.*, 1999; Sicardy *et al.*, 1999) show that the exact ring location is outside the predicted location from Goldreich's model at the 42:43 Galatea resonance. Fortunately, a different resonance may be able to provide the confinement (Namouni and Porco, 2002). Thus, the cause of the extreme azimuthal structure observed in Neptune's Adams ring is still not clear and quite unlikely to be a remnant of the formation of Neptune. It is more likely to be the result of recent catastrophic events.

Table 10.1 *Estimated satellite disruption lifetimes for two strength models*

			Disruption lifetime (10^9 years)	
Satellite	Radius (km)	Orbit radius (10^4 km)	Durda *et al.* (1998)	Housen *et al.* (1991)
Uranus				
Puck	77	8.60	56	4.0
Belinda	34	7.53	9.1	1.3
Rosalind	29	6.99	5.9	1.0
Portia	55	6.61	17	1.8
Juliet	42	6.44	10	1.3
Desdemona	29	6.27	5.0	0.83
Cressida	33	6.18	6.2	0.91
Bianca	22	5.92	2.9	0.59
Ophelia	16	5.38	1.4	0.38
Cordelia	13	4.98	0.91	0.29
Neptune				
Proteus	208	11.76	1.2×10^4	25
Larissa	96	7.35	83	3.7
Galatea	79	6.19	29	2.3
Despina	74	5.25	20	1.7
Thalassa	40	5.01	5.9	0.83
Naiad	29	4.82	3.0	0.56

Source: Colwell *et al.* (2000).

10.7 Satellite disruption

Making rings by disrupting satellites naturally leads to consideration of the moons' evolution. The history of the small satellites of Uranus and Neptune has been studied in detail by Colwell and Esposito (1992, 1993) and Colwell *et al.* (2000). Their stochastic simulations of the moons' collisional fragmentation confirm the conclusions of Smith *et al.* (1986, 1989) that these moons are not primordial. Table 10.1 gives estimated disruption lifetimes for small satellites of Uranus and Neptune. Many of these are significantly less than the age of the solar system. Different models for the moons' strength in Table 10.1 by Durda *et al.* (1998) and Housen and Holsapple (1990) provide seriously different lifetimes (Colwell *et al.*, 2000), which do not allow us to establish a definitive history but instead give the general outline of destruction of small moons to make rings. Colwell and Esposito (1992, 1993) follow the process of satellite disruption from an initial distribution through successive disruptions, ignoring reaccretion. Two approaches are used: a Monte Carlo simulation following the history of only the largest fragment after each disruption, and a Markov chain calculation following the stochastic evolution of all the fragments (see Chapter 9 for more discussion of this approach).

These simulations raise the question of reaccretion of fragments of a disrupted satellite in or near the planet's Roche zone. Previous estimates by Burns *et al.* (1984) and David Stevenson *et al.* (1986) are based on the calculation of Steven Soter (1971), who computed the timescale for an ejected particle to collide with its source body. Canup and Esposito (1995) have used a matrix formulation of the integro-differential *coagulation equation* and applied it to the ensemble of satellite fragments calculated by Colwell and Esposito (1992), to improve on Soter's (1971) calculation. This method was a particle-in-a-box simulation of the mutual collision statistics. Their results show that reaccretion cannot be ignored: realistic calculations in the Roche zone must form a critical part of our understanding of ring and moon evolution. Already we expect some limited reaccretion within rings as particles may temporarily stick together between higher velocity collisions. These temporary aggregations were termed dynamic ephemeral bodies (DEBs) by Stuart Weidenschilling *et al.* (1984) (see Figure 4.6). Despite this reaccretion, the existence of rings surrounding the planets shows that accretion does not dominate near the planet, and the present models' general neglect of accretion in ring-history calculations seems a justifiable simplification, but the long-term evolution may require some balancing accretion, see Section 10.9.

Once the original moon is destroyed for the first time, the collisional cascade to smaller size through successive disruptions occurs relatively quickly, since the smaller moons have less gravitational binding energy, and are easier to destroy. Monte Carlo simulations of 300 separate histories of the moons of Neptune give the results shown in Figure 10.1. For an original moon of radius 40 km, the solid line shows the mean, asterisks the median, and squares the mode for the distribution of the largest surviving fragment as a function of time. These results show that the evolution of moon populations from catastrophic fragmentation is more complex than can be described by a simple timescale. The speed of this cascade is further dependent on the size distribution of the impactors, which is, unfortunately, a poorly known factor. For example, the difference between a power-law index of 2.5 and 3.5 is quite significant. Better-constrained measurements of the impactor size distribution are strongly desired, although some recent values do not change our conclusions significantly (Colwell *et al.*, 2000).

The Markov chain method used for these calculations (see Chapter 9) follows the evolution of the complete size distribution of moons and fragments. These simulations show that the more numerous, smaller fragments can outlive the largest fragment followed in the Monte Carlo formulation, yielding collisional debris from an initial complement of satellites that has an approximate power-law distribution. Colwell and Esposito (1992, 1993) find a cumulative-size power-law index of

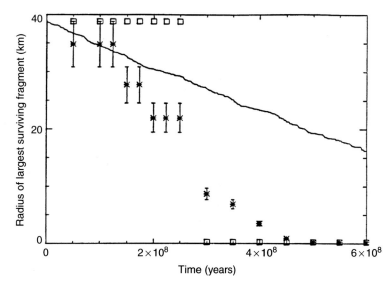

Figure 10.1 Mean (–), median (*), and modal (□) sizes of a distribution of 300 stochastic simulations of the collisional cascade for an initial 40-km-radius moon, at the orbit of Thalassa. Error bars on the median connect the two size bins, which fall on either side of the 50% cumulative probability. (From Colwell and Esposito, 1992.)

2.54 for the collisionally evolved system. We note that, after 4 billion years, the remnants of an original 100-km-radius moon would include about one thousand 1-to-10-km moonlets.

The short lifetimes calculated by Colwell and Esposito (1992) and Colwell *et al.* (2000) show that the small satellites of Neptune must have evolved through catastrophic fragmentation since the end of satellite and planet formation 4 billion years ago. The production of the currently observed smaller satellites may be a natural consequence of the successive break-up of larger satellites.

10.8 Ring formation

Through further study of the size distribution and velocity distribution of the satellite fragments, Colwell and Esposito (1993) found that a narrow ring is the natural outcome of a disruption, as the fragments' orbits quickly fill the entire circumference at the original orbital radius of the destroyed moon. In the production of this debris, a small number of large fragments are also created. Colwell and Esposito (1990a) argue that these larger fragments will naturally clear gaps. The discovery of Pan in Saturn's Encke Gap (Showalter, 1991) and Daphnis in the Keeler Gap (Porco *et al.*, 2005) provide examples of this. As the initial ring spreads, its edges will cross through the resonances with larger bodies in the system, allowing the edges of the rings to be shepherded and sharpened. The moons will then spread with the rings (albeit more slowly because of the moons' larger mass), and possibly their evolving orbits will resonate with yet larger satellites. A natural result is that the evolved system will include many mutual resonances. This can explain the "Cordelia connection" found by Murray and Thompson (1990) for the Uranus rings, that is, that several of the unseen moons that are hypothesized to halt the radial spreading of the Uranus rings are exceptionally close to resonances with Cordelia. Furthermore, they noted that Cordelia itself is very close to a resonance with Rosalind. This linking-up, due to orbital resonances and associated transfer of angular momentum successively from smaller to more massive satellites, will slow the initially rapid viscous spreading of the newly formed ring.

The moons can also confine the ring material in *azimuth*: Neptune's arcs provide a vivid example (see Chapter 13). Like the radial lock-up, this confinement occurs quite quickly; otherwise, differential rotation due to Kepler shear and differential precession would smear out the arcs on a timescale of years. Porco (1991) has shown how the azimuthal structure of the ring arcs may be understood as material trapped in 7 of 86 possible Galatea corotation resonances (see Figure 10.2). All the ring arcs span an azimuthal range of only 25°; thus, the resonance locations are only thinly occupied.

We can estimate the probability that Neptune's particular arrangement arose by a set of chance events (Esposito and Colwell, 1992). One possibility is that each of the resonance locations was filled by an independent event. Examples might be a collision between two fast-moving particles in one of the resonance zones, or the destruction of a small moon at or near the resonance site. Each event would release material to be subsequently trapped in that corotation resonance.

Esposito and Colwell find that the random hypothesis is excluded with a likelihood 1×10^{-8} (i.e., the probability of chance occurrence of all the ring arcs clustering together is 10^{-8}). The conclusion from this calculation is that we are most likely seeing the result of a single recent event. This could have been the collision of a meteoroid with a small moon of Neptune, resulting in the destruction of the moon; the smaller fragments would now be trapped in 7 of the 10 nearby corotation sites. This destruction could also simultaneously leave several large fragments (as explained above) to assist in the azimuthal confinement, as proposed in the model of Jack Lissauer and Bruno Sicardy (1990).

The short lifetimes inferred from meteoroid bombardment imply interrelated histories for rings and nearby small moons in the planetary ring systems. Detailed simulations show the plausibility of a model where disrupted satellites provide the source of the Uranus and Neptune rings. The satellites confine the rings radially and azimuthally. The entire ring–moon system may lock up through resonances

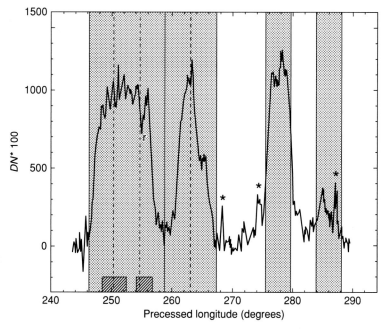

Figure 10.2 A radially averaged longitudinal scan of the Neptune rings arc region precessed back to a common epoch using the arcs' mean motion, $n = 820.1185°$ per day as determined by the Voyager observations. The asterisks indicate the positions of stars; the symbol r indicates an incompletely removed camera marks. The stippled bars show the location of corotation resonances, each 4.186° wide, which might explain the azimuthal confinement of the observed arcs in the model of Goldreich *et al.* (1986). As can be seen, not all resonance zones are occupied by arcs, and some of the observed arcs fill more than one resonance. (Reprinted with permission from Porco, C.C., *Science*, **253**, 995–1001, © 1991 American Association for the Advancement of Science.)

and evolve as a unit. Satellites excite the random motions within the rings. The resulting interparticle velocities determine the size distribution through collisional equilibrium. The collisions also produce the significant amounts of dust observed by the Voyager cameras in the Uranus and Neptune rings.

10.9 Cosmic recycling

We have now seen that planetary rings appear much younger than the solar system. Many processes act quickly on ring systems. Mutual collisions and meteoroid bombardment grind the ring particles while charged particles sputter molecules from their surfaces. Unless confined, the rings spread as their particles exchange momentum due to collisions and gravitational scattering. Even if confined by shepherding satellites, this only slows the process: the momentum is instead transferred to the shepherding moons via the resonance at the ring's edge. Because the moon is more massive than the ring, its evolution is much slower, but nonetheless it steadily moves away from the ring due to conservation of angular momentum. The abundant density waves in the rings also transfer momentum to the perturbing moons. They likewise recede. As an example, tiny Atlas, which orbits just outside Saturn's A ring, would have evolved to its present location in less than 10 million years. Similar short timescales are found for Prometheus and Pandora, the F-ring shepherds. In Section 10.5, we estimated a maximum lifetime of the Uranian ε-ring shepherds of less than 600 million years.

A similar constraint is provided by the meteoritic pollution of Saturn's rings. The constant rain of meteoroids onto the rings brings dark carbonaceous material, which would quickly darken their bright, icy surfaces in less than 300 million years, as described by Jeff Cuzzi (1998) of NASA's Ames Research Center.

These calculations give the lifetime of the current rings and moons against the various processes ranging from 10^7 to 10^9 years: at the current rates they would now be all gone, unless they had been created more recently. As discussed above, a potential solution is provided by the fact that we find rings and moons intermixed in all the ring systems. Voyager, Galileo and Cassini show clear ring–moon interactions. Moons sculpt, sweep up, and release ring material. Detailed calculations by Canup and Esposito (1995) and Barbara and Esposito (2002) show that this distribution of both small and large bodies coexisting is a natural outcome of their proximity to the planet, near the Roche limit, in a region termed the Roche zone (see Section 4.4). Too close to the planet, its tidal forces dominate over the mutual gravitational attraction between orbiting bodies and they cannot stick together from the force of their own gravity. Far from the planet, its gravity is small compared with the mutual attraction of two bodies; they can hold together and even collect more objects due to their mutual attraction. In between, the Roche zone is where the competition between these phenomena is more even. Not all colliding bodies in the Roche zone will stick together: close to the planet, two like-sized objects cannot be held together by their own gravity. On the other hand, a small body can be held by the gravity on the surface of one that is much larger. This uneven behavior leads naturally to a bi-modal distribution. The largest bodies persist because they are too close in size to accrete each other; they collect the smaller bodies into a regolith of fragments and dust that covers the particle's surface. Occasional collisions liberate this regolith which orbits freely until recaptured by another large body. Thus, an equilibrium is established that contains both large and small ring particles. The largest may be big enough to be termed moons. This explains the coexistence of rings and moons.

Esposito and Colwell hypothesized that these moons are the progenitors for future rings. A moon shattered by a large impact from an interplanetary projectile

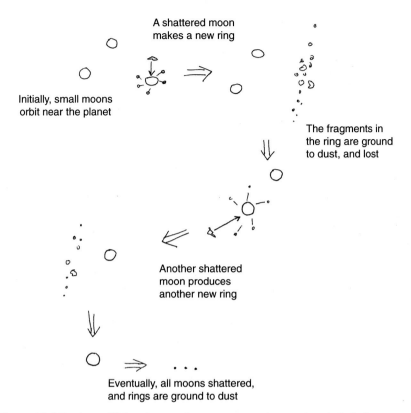

Figure 10.3 In the collisional cascade, moons are shattered and their fragments further broken to make rings and dust. Eventually, the last moon is destroyed and the original material is completely ground to dust. (From Esposito and Colwell, 2003.)

would become a ring of material orbiting the planet. Big moons are the source of small moons, small moons are the source for rings. Rings are eventually ground to dust that is lost by becoming charged and carried away by the planet's rotating magnetic field or by atmospheric drag into the planetary atmosphere (where it shines briefly as a meteor). We called this process a collisional cascade (see Figure 10.3).

The largest fragments could shepherd the smaller ring particles. Rings and moons would spread together as a unit, linked by mutual resonances. Small moons will be caught in resonance with larger moons: this slows the linked evolution. This general story explains the coexistence of rings and moons, the recent origin of rings and moons from recently shattered moons, and the size distribution of ring particles in a ring. Since the shattering of a moon is a random event, ring history will be stochastic and somewhat unpredictable. The differences between the various ring systems are explained by the different random outcomes of this stochastic process. Thus, the collisional cascade can provide an explanation for the apparently different ring systems around each of the giant planets. Catastrophic events provide the tempo for creating planetary rings: new rings are episodically created by destruction of small moons near the planet. This disorderly history arises from singular events. A ring's physical nature is determined by the balance of competition between fragmentation and accretion in the planet's Roche zone.

This hypothesis also has some problems. The collisional cascade uses the raw material (a planet's initial complement of moons) too rapidly. If we imagine we are now looking at the remnants of 4.5 billion years of successive destruction over the age of the solar system, then this process is almost at its end. The small moons that now remain as the source of future rings have a lifetime of only some few hundred million years, based on calculations by Colwell *et al.* (2000). This is less than 10% of the age of the solar system. Why are we humans so fortunate as to come upon the scene with robotic space exploration, just in time to see the ring finale?

A second problem for the collisional cascade model is with explaining the recent origin of Saturn's rings. Saturn has the most massive ring system, and its multitude of rings span the entire variety of ring phenomena seen around all the giant planets. The total mass of Saturn's rings has been estimated from observing density waves propagating through the ring system. These waves provide a remote diagnostic of the physical properties of the ring system. The analysis by Linda Spilker of the Jet Propulsion Laboratory and her colleagues (Spilker *et al.*, 2004) confirms earlier estimates that the Saturn ring system has a mass about equivalent to the mass of the moon Mimas, which is 200 km in radius. It is a very unlikely event that a previous moon of this size was destroyed by an impact in the last few hundred million years. Although it is easy to imagine the origin of Saturn's rings as the result of a random occurrence, one so rare as this causes a fundamental problem. More recent calculations by Robbins *et al.* (2010) show that occultations underestimate the total ring mass, especially in denser rings. My own reanalysis of Pioneer 11 results (in Charnoz *et al.*, 2009) for the mass of ring B considers an alternative interpretation of these results. The small number of neutrons emerging from Saturn's rings could be the result of either a low-mass ring or of a high-mass ring that itself absorbs some of the neutrons created by cosmic rays hitting the rings. In that case, Saturn's B ring would be more than five times as massive as simple extrapolation of the surface mass density from observed density waves would predict. We can not check this possibility with current information from density waves, because density waves are not observed in the rings' thickest parts.

As an alternative, Luke Dones of the Southwest Research Institute of Boulder, Colorado, has proposed that a large comet, the size of the object Chiron, passed close enough to Saturn that the tidal forces disrupted it. The fragments of the comet Shoemaker–Levy 9 were similarly created when it passed too close to Jupiter,

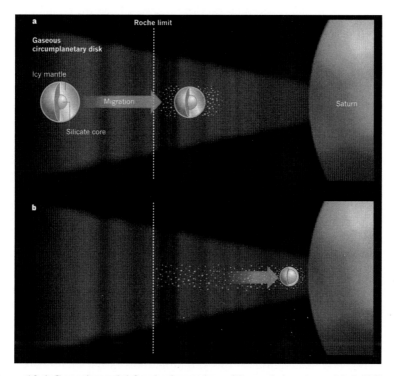

Figure 10.4 Canup's model for the formation of Saturn's icy rings. (a) A differentiated satellite in the gaseous circumplanetary disk around Saturn migrates towards the planet. When the satellite crosses the planet's Roche limit, its icy mantle starts to be pulled into pieces by the planet's tidal forces. (b) The silicate core carries on migrating towards Saturn and eventually falls into it, leaving behind the ice which gives birth to the rings and icy satellites of the planet. (From Crida and Charnoz (2010); reprinted with permission from *Nature*.) For color version, see Plates section.

well inside its Roche limit. These fragments did not form a ring, but spectacularly collided with Jupiter in 1994. Needless to say, Dones' calculations show that such a disruption is also an exceptionally rare event.

An exciting new proposal by Robin Canup of the Southwest Research Institute in Boulder is that the rings of Saturn may have formed from a Titan-sized moon whose icy upper layers were ripped off as it spiraled into Saturn shortly after the planet formed (see Figure 10.4). This would explain the rings being nearly 100% water ice. The present moons of Saturn are all about half ice, half rock. Furthermore, meteoritic pollution by interplanetary dust would pollute and darken any primordial icy rings, making their present icy composition difficult to explain.

If the initial satellite in Canup's model had already differentiated, its rock would be concentrated in a central core, surrounded by a mantle of pure water ice. Because of the friction from the dusty disk that the moons of Saturn formed in, they are slowly drifting inward toward the planet. Tidal forces can disrupt the icy mantle, leaving the rocky core intact to be dragged into Saturn and lost. This leaves a pure ice ring, which can itself recondense into new moons that spiral outward as the disk dissipates. Canup's theory starts with a much more massive ring, gradually spreading, truncated by resonances with Saturn's more distant moons and spawning new, icy satellites at its outer edge. Cassini will be able to test this at the end of its mission by measuring the ring's mass.

These problems of explaining the persistence of rings and moons in planetary ring systems, and particularly of Saturn's massive rings, can be addressed by a careful consideration of recycling the raw material. The early calculations of ring–moon evolution considered a single initial moon in isolation: non-linear and collective effects were ignored in the original papers by Colwell and Esposito. Further, reaccretion was not considered. Canup and Esposito (1995) calculated evolution of interacting bodies within the Roche zone, where accretion is possible if the mass ratio of the colliding bodies is large enough and the collision is sufficiently dissipative. Thus, particles knocked off or resulting from the disruption of one parent body can be recaptured by another of the moons residing in the ring system. The entire system achieves a balance between the competing processes of fragmentation and reaccretion. In this situation, most of the small material resides predominantly on the surfaces of the larger bodies. It is protected from meteoroid bombardment, sputtering, and darkening by the overburden of regolith above it. Only the top layer of regolith material is subject to these destructive processes. The larger parent bodies thus would resemble piles of rubble. These unconsolidated objects can be disrupted, accrete again (or onto other bodies), continually resupplying the material to make rings. In order to maintain the purity of the initial ice, the rings must contain about 10 times more mass than estimated from the density waves (see Elliott and Esposito, 2011).

Just how much difference can recycling make? A simple probabilistic calculation allows us to estimate its effectiveness. We use a Markov chain model (see Chapter 9) to simulate the recycling of ring material. In this model, we consider a small element of mass participating in the collisional cascade. The discrete states of the Markov chain represent the size of the body on which this element resides. We imagine $n+1$ states, each representing a size bin, together spanning the size range from the initial moon (10 km or more in radius) to the finest dust, less than 1 μm in size. If this size range of 10^{10} is divided into logarithmically spaced bins, each twice as large as the next, we would have approximately $n \approx 30$ states of the system. The transition matrix for this Markov chain gives the probability of going from one state to the next (changing the size of the body that the mass element lies in) as

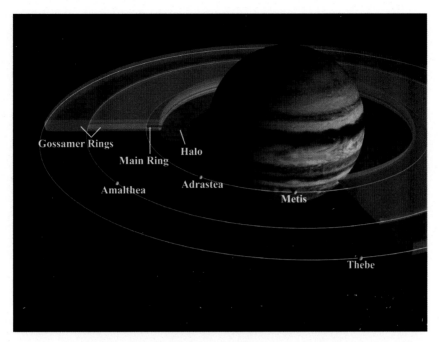

Figure 3.3 For full caption please see text (p. 25).

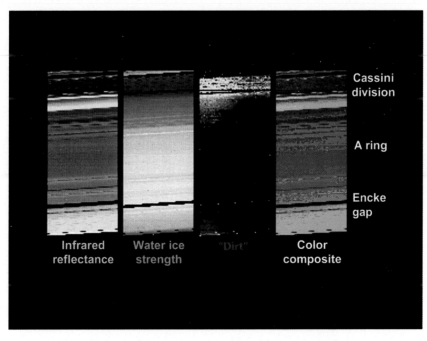

Figure 4.8 For full caption please see text (p. 46).

Figure 7.3 For full caption please see text (p. 72).

Figure 7.4 For full caption please see text (p. 72).

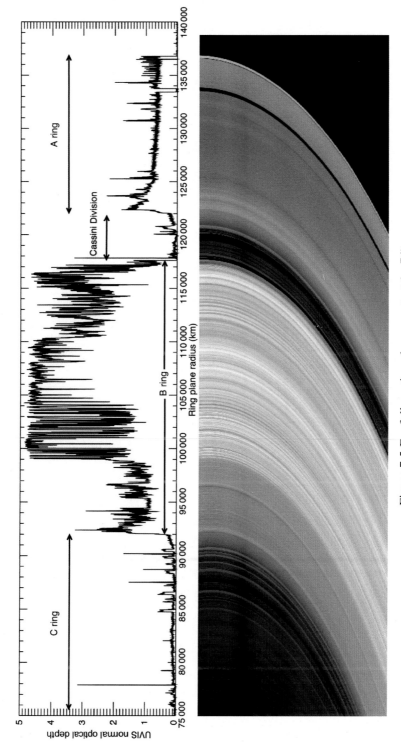

Figure 7.5 For full caption please see text (p. 74).

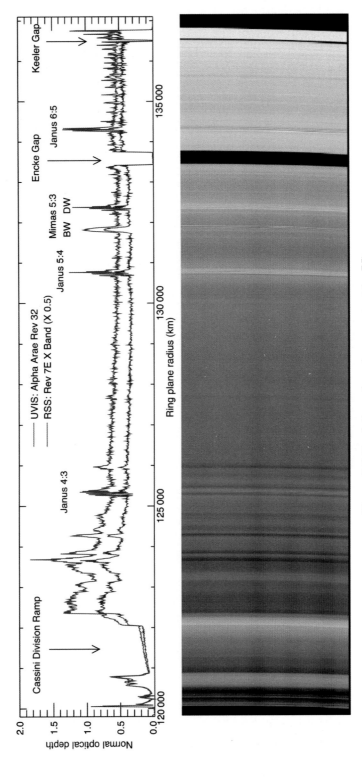

Figure 7.6 For full caption please see text (p. 75).

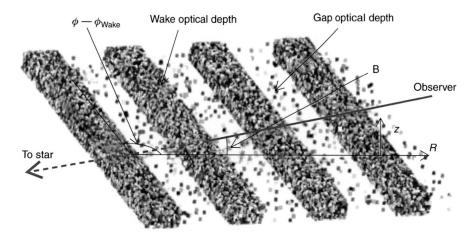

Figure 7.9 For full caption please see text (p. 78).

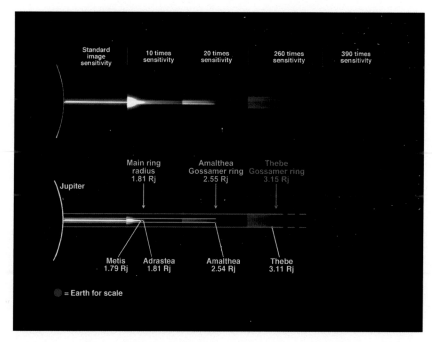

Figure 7.23 For full caption please see text (p. 94).

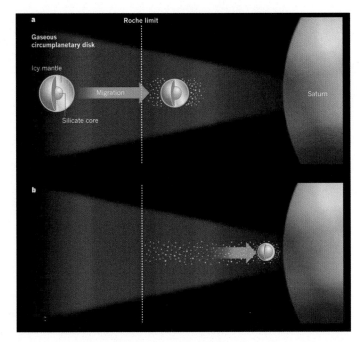

Figure 10.4 For full caption please see text (p. 129).

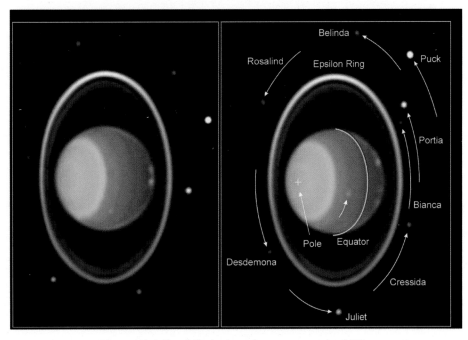

Figure 12.1 For full caption please see text (p. 152).

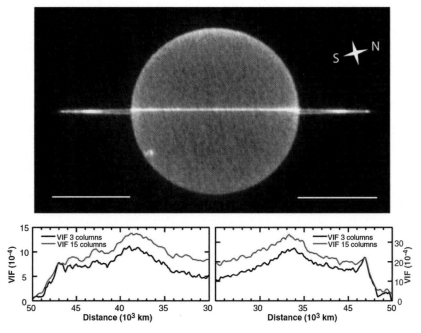

Figure 12.2 For full caption please see text (p. 153).

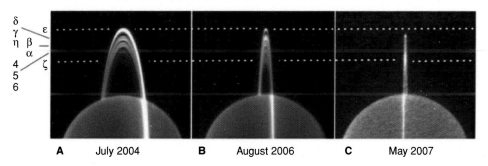

A July 2004	**B** August 2006	**C** May 2007

Figure 12.3 For full caption please see text (p. 154).

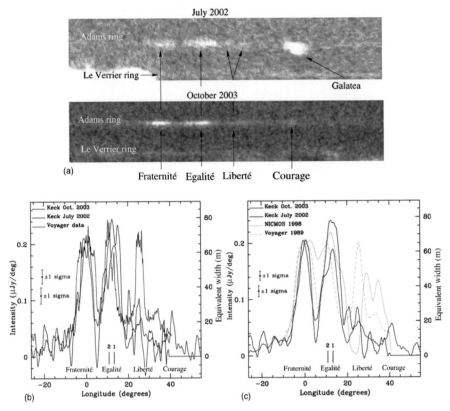

July 2002

Adams ring

Le Verrier ring

Galatea

October 2003

Adams ring

Le Verrier ring

(a)

Fraternité Egalité Liberté Courage

(b)

(c)

Figure 13.8 For full caption please see text (p. 169).

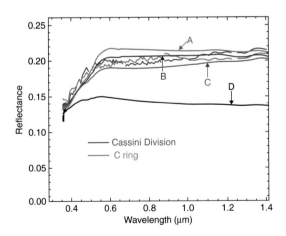

Figure 15.2 For full caption please see text (p. 185).

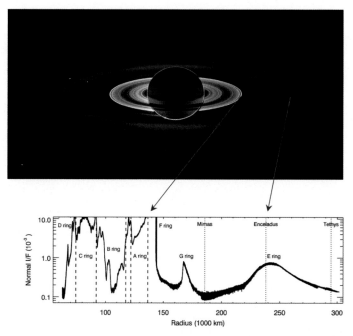

Figure 16.4 For full caption please see text (p. 195).

Figure A-2 For full caption please see text (p. 210).

Figure A-1 For full caption please see text (p. 209).

Figure A-3 For full caption please see text (p. 211).

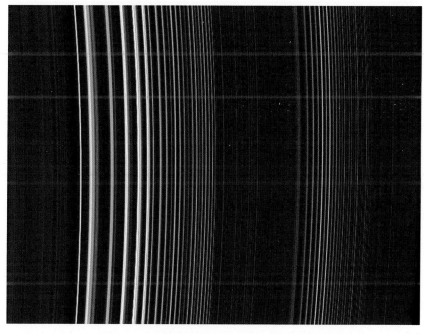

Figure A-4 For full caption please see text (p. 211).

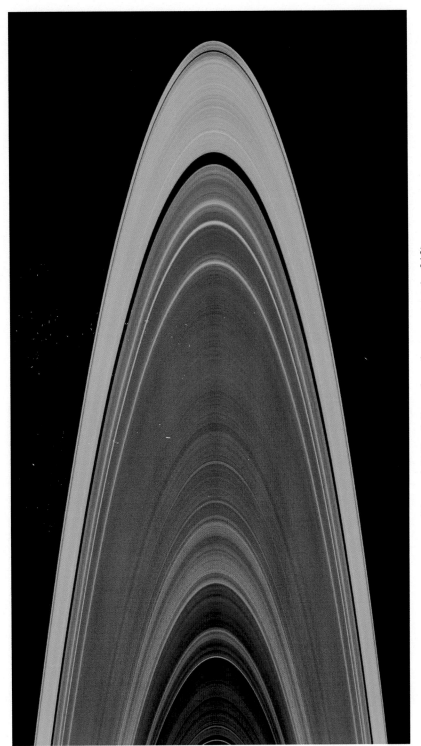

Figure A-5 For full caption please see text (p. 212).

Figure A-7 For full caption please see text (p. 213).

Figure A-8 For full caption please see text (p. 214).

Figure A-10 For full caption please see text (p. 215).

Figure A-12 For full caption please see text (p. 216).

Figure A-13 For full caption please see text (p. 217).

the elements of a square $(n+1) \times (n+1)$ matrix. We will see below that, in our model of the matrix, most elements vanish. For definiteness, we can label the initial largest body as state 0. The ephemeral dust is in state n. After every disruption event, the chain advances to a higher state; conversely, after accretion the index of the current state decreases.

Our collisional cascade has n steps, from moons to dust to gone (free-floating dust is rapidly lost due to plasma drag, *Poynting–Robertson drag*, and sputtering by trapped radiation; see Section 16.1). In terms of our earliest models, which neglected recycling, we see that, after n steps of the stochastic evolution, the original ring material would be completely gone. From our understanding of the flux of large bodies passing near the outer planets, we can estimate a lifetime of 100 million years for a moon of size $R = 15$ km (see Colwell *et al.*, 2000). Even starting with an initial complement of 100 moons, we would expect to have only one moon remaining after 500 million years. The probability of a moon this size surviving for the age of the solar system is 10^{-18}. Because larger bodies have longer lifetimes, we can lengthen the timescales if we store the materials in larger moons. These, in turn, would be disrupted more recently to produce the present ring progenitors. Nonetheless, this collisional cascade leads us to the philosophically unsatisfying conclusion that, now, nearly all the raw material for rings around the planets is gone. This is, of course, *without* recycling.

To calculate the effect of recycling, we define the following properties of our Markov chain model. With probability p, the system will advance to the next state, $j \rightarrow j+1$ (this is a disruption, where the typical largest remaining fragment is half the mass of the original object). With probability q, the system returns to the initial state, $j = 0$ (the element of mass we are considering is swept up by a moon). Naturally, we have $p + q = 1$. Note that, for simplicity, we take the lifetime of each bin constant and we consider only sweep-up by the largest parent body. This allows us to estimate the lifetime of an initial moon, including recycling. The probability q represents the probability that a fragment is swept up before its next disruption in the collisional cascade. In our first simulations (see Colwell and Esposito, 1992), this probability was implicitly set at the value $q = 0$.

The chain I have described is classified as an absorbing Markov chain, with one absorbing state, $j = n$. The other states are all transient. We estimate the ring–moon lifetime as the mean time to absorption (measured in steps), with the chain starting from the initial state, $j = 0$.

With p as the probability of fragmentation and q as the probability of reaccretion, we calculate the lifetime of the initial moon as

$$E_0 = (1 - p^n)/(p^n q) \qquad (10.5)$$

This solution comes from writing the set of expected times to absorption as the vector $\vec{E}$, where E_j is the mean time for state j. If the transition matrix is $\overset{\leftrightarrow}{P}$, then we must solve the matrix equation

$$\vec{E} = 1 + \overset{\leftrightarrow}{P} \cdot \vec{E} \tag{10.6}$$

to determine $\vec{E}$. This equation relates the absorption times now and one time-step later. Note that nearly all the elements of P are zero: this makes the solution straightforward by back-substitution (see, e.g., Baldi *et al.*, 2002, p. 124).

   Although this is a highly idealized model, the effect of recycling can be clearly seen by varying the value of q.

(1) First, let q be small, that is, $nq \ll 1$. This is the case for negligible recycling.

$$E_0 = (1 - (1 - q)^n)/((1 - q)^n q) \tag{10.7}$$

Then using the binomial theorem $(1-q)^n \approx 1 - nq$, we find

$$\begin{aligned} E_0 &\approx (1 - 1 + nq)/(q(1 - nq)) \\ &\approx nq/q \\ &\approx n \end{aligned} \tag{10.8}$$

The lifetime is linear: it is just proportional to the number of states. This reproduces the familiar result from the collisional cascade without recycling.

(2) Second, consider $nq \to 1$. The same approximation yields

$$\begin{aligned} E_0 &\approx n(1 + nq) \\ &\approx n^2 q \end{aligned} \tag{10.9}$$

The lifetime varies as n^2; this is similar to diffusion, where it takes n^2 steps to reach a distance n from the origin.

(3) For $p = q = 1/2$, we get

$$E_0 = \frac{1 - \left(\frac{1}{2}\right)^n}{\left(\frac{1}{2}\right)^{n+1}} \approx 2^{n+1} \tag{10.10}$$

We note that recycling 50% of the material at each step lengthens the expected lifetime from n to 2^n steps!

(4) Suppose ring systems recycled most of the fragments. This would be the case if other moons recapture the majority of fragments. Then

$$q \to 1 \text{ and } E_0 \to p^{-n} \text{ as } q \to 1 \tag{10.11}$$

For example, for $q = 0.9$, we estimate $E_0 \sim 10^n$. This gives an indefinitely long lifetime for q close enough to unity. Calculations by Barbara and Esposito (2002) show that, at equilibrium after the disruption of a small moon, about 50% of the mass of small material is collected by the larger bodies. Throop and Esposito (1998) estimated that only about 10 g per year is lost from Saturn's G ring. This current loss rate implies no significant loss of mass from that ring over the age of the solar system. These calculations show clearly that recycling can extend the age of rings indefinitely. For Saturn's rings, this would remove the requirement to create them in the recent past; instead they would have been continually changing and recycling over the past 4.5 billion years. Although the rings we see now are recent, the *ring system* is continually renewed and probably ancient. Using an urban analogy, a city can persist for millennia, even though its individual citizens live but three score years and ten.

This recycling hypothesis provides answers to the questions of the origin of Saturn's rings; we need not worry that we are now at the apparent end of the age of rings. The concern about the meteoroid darkening of rings can also be removed, all by invoking cosmic recycling. This recycling will bury the meteoritic infall, diluting it to the small fractions inferred by Cassini. Although individual rings and moons are ephemeral, ring–moon systems persist. For Saturn, this would mean that the B ring is ancient, while the other rings are all ephemeral. Ring systems will go through a long quasi-static stage where their optical depth and number of parent bodies slowly decline. Nonetheless, the overall mass of the rings would be eventually depleted. For material to last the age of the solar system requires the rings be about 10 times more massive than calculated from density waves (see Esposito *et al.*, 1983b), or that the flux of impactors onto the ring is 10 times less, or some combination (Elliot and Esposito, 2011; Esposito *et al.*, 2012). When just a few parent bodies remain, recycling becomes less effective and the collisional cascade proceeds rapidly to its conclusion.

This hypothesis does not answer all the questions raised about the apparent youth of planetary rings. Some rings are too close to their planet for recapture to be likely. For Uranus' and Neptune's rings, we may need to better understand the dynamics of break-up of a rubble pile. Perhaps the strength of such unconsolidated bodies is much more uniform as a function of size, as discussed by Colwell *et al.* (2000). This behavior could also serve to slow significantly the collisional cascade. Momentum transfers between rings and moons are still a problem: we would still expect to see rapidly spreading rings and rapidly receding moons. However, the recent discovery of the chaotic motions of Saturn's moons Pandora and Prometheus may indicate that the resonant interactions between moons is more complicated than we previously thought. Perhaps temporary resonance captures can serve to transfer momentum to larger moons, like Saturn's moon Mimas. Pandora is now just slightly out of resonance with Mimas. If it has had significant resonant interactions with that larger moon, then the timescale needed to reach its current position, and those timescales estimated for other parts of Saturn's rings that may have interacted with Pandora in the past, are possibly greatly underestimated. See discussions by Borderies *et al.* (1985), French *et al.* (2003), and Goldreich and Rappaport (2003).

The current opportunity for progress on the issue of the role of *cosmic recycling* in ring history is provided by the close-up views of Saturn's rings from Cassini. Cassini has probed the current balance between accretion and fragmentation in the F ring; it has constrained the particle size distribution and composition of the ring particles; it has allowed us to compare theoretical models of ring dynamics to measurements in order to test our explanations of current ring behavior. We have not yet confirmed the hypothesis that the continuing fragmentation and re-accretion balance each other to create the present rings; that the color and scattering properties of the ring particles are due to material recaptured by them; and that these processes are continually renewed by recycling of material between rings and moons. A major open question of the total mass of Saturn's rings can be answered by Cassini, near the end of its mission in 2017. The gravitational pull from the rings on the spacecraft will determine the ring mass accurately.

10.10 Summary

Processes in planetary rings are proceeding so rapidly that many structures we observe cannot be primordial. Cassini sees significant changes since Voyager, and even since the start of the Cassini mission (French *et al.*, 2012). These include narrow rings, dusty rings, sharp edges, Saturn's bright A and B rings, and Neptune's partial rings. We believe rings are the likely result of shattering of some former moon that orbited the planet. However, the situation is not quite so simple as a collisional cascade, with large moons broken to small moons, broken to rings, and then ground to dust. Moons sculpt, sweep up, and release ring material. The moons we see now are probably piles of rubble, regathered from previously shattered moons. Thus, planetary ring systems show clear evidence of recycling, where primordial material is used again and again. The reaccretion of ring material provides sources that can continually resupply material to make new rings. Calculations show that, if the recycling is large enough, ring lifetimes can be indefinitely extended. In that explanation, the rings continually change, the current structures cannot persist, but the ring *system* is continually renewed. Cassini observations can confirm and refine this recycling hypothesis, but have not yet provided decisive evidence.

11

Saturn's mysterious F ring

In September 1979, I was at the NASA Ames Research Center in Mountain View, California. The aging spacecraft Pioneer 11, launched in 1973, was making its long-awaited approach to the planet Saturn. It was the first spacecraft to reach that planet, although Voyager 1 and Voyager 2 had been launched on a shorter route two years earlier in 1977 and were now speeding toward the ringed planet, to fly by in 1980 and 1981. The team leader for Pioneer 11's unsophisticated camera, Tom Gehrels, had invited me to join his team, my responsibility being to cover Saturn's ring system. The original target of Pioneer 11 had been Jupiter, whose gravity had redirected the spacecraft toward its looming encounter with Saturn. Now, Gehrels was expanding the imaging team to include expertise needed for Saturn. My office-mate Bob West had informed his former thesis advisor, Marty Tomasko (who was at the University of Arizona, Tucson, along with Gehrels, and a key member of the Pioneer team), that the standard methods of *radiative transfer* (see Chapter 15) they had used to analyze the atmosphere of Jupiter might not be adequate for Saturn's rings, where the individual particles cast shadows on each other, unlike the molecules in a gas. Thus, I received a call to join the team. My tasks were to serve as a general resource for any ring studies and to apply specialized techniques I had developed in my Ph.D. dissertation.

My assignment, in particular, was to be on the lookout for new rings. When Pioneer 11 was launched, Saturn was the only planet known to have rings. However, while it was on its way to Saturn, nine narrow, dark rings had been discovered around the planet Uranus and a diffuse, dim ring around Jupiter had been captured in stunning photos from the Voyager spacecraft earlier in the year 1979. The possibility of new rings was on the minds of scientists and the public. Listening to the radio broadcasts from the Ames Research Center before my arrival, I had noted that there were a number of questions from the press about rings. It was a slight worry to me that my colleagues were deferring these questions until my arrival, which I had planned shortly before Pioneer's closest approach.

The Pioneer images were few and grainy. The rings were backlit at closest approach, an aspect never visible from the Earth. Because Pioneer viewed the rings from the side opposite the Sun, a partial contrast reversal occurred: the denser parts of the ring system that usually appeared bright because they reflected sunlight toward the Earth were now opaque and dark. In the Pioneer images, the thinner parts of the rings were most obvious, lit up like a dark cloud's "silver lining." The grainy images and unusual aspect were a challenge to interpreting the close-up information. I retired to my cubicle with pencil and graph paper to try to make sense of these data. By dawn on that September day, I had convinced myself that a new narrow ring was outside the known main rings of Saturn. I was able to convince Gehrels, also, which allowed this discovery to be announced later that day at a press conference, where I painstakingly pointed to lines and tiny numbers on several pieces of graph paper tacked to an easel.

Because a number of new rings had been reported and contested in the previous decade, Gehrels was concerned both about the certainty of our results and about their nomenclature. New rings with some letter designations up to "Z" had been claimed. Gehrels and I had been skeptical of some of these reports before; our Pioneer pictures showed a number of these reports to be erroneous, and we did not wish to confuse our colleagues and the public. At the same time, we realized that more rings might still be discovered later (that were beyond Pioneer's capability to detect) in the same regions where the erroneous observations had been claimed. We did not yet recognize the multiplicity and abundance of rings soon to be revealed by the Voyager missions, but we wished to be cautious and avoid confusion.

Thus, this new narrow ring was not named ring "E" (which could have denoted "Esposito") or even ring "D," since our pictures clearly denied the previous statements about such rings from some ground-based observations. Proceeding alphabetically, we called the new ring "F," later affirmed officially by the International Astronomical Union's working group on nomenclature.

In addition to the heady excitement of spacecraft encounters and press attention, another aspect of my experience was weighing on my mind. The simple model of planetary rings as uniform and unchanging that I had grown comfortable with in my own dissertation on Saturn's rings (1978) was unsatisfactory. Even the under-resolved images from Pioneer were incompatible with this simple model. I wrote in the team paper in *Science* magazine: "These new results mean that the ring system is more complicated than expected from Earth... A placid, homogeneous model of the rings is not consistent with our data" (Gehrels *et al.*, 1980).

This sense of unresolved complexity did not prepare us for the Voyager images that arrived 14 months later in 1980. The F ring was seen in great detail by the Voyager cameras. These pictures show multiple strands, kinks, clumps, and even "braids" where the strands appear to cross. Imaging team leader Brad Smith opined in a press conference that the pictures defied the laws of physics; Peter

Goldreich later that day commented, "Don't worry about Newton's equations; most people don't realize how many solutions they have" (Elliot and Kerr, 1984). We are still working today to try to explain the F-ring features seen by Voyager (Showalter, 1998; Barbara and Esposito, 2002), and the more recent Cassini images and occultations (Albers *et al.*, 2012; French *et al.*, 2012). In retrospect, it is not at all surprising that the newest findings show even more dynamic structures!

11.1 F-ring structure

The most detailed pre-Cassini analysis of the structure of the ring (Murray *et al.*, 1997) showed that the F ring is composed of four strands that do not interact. The original Voyager statements about braiding were thus overstated. All of these strands extend over at least 45° in longitude. Features change on a timescale of months, but the strand orbital elements did not change significantly between the Voyager 1 encounter in 1980 and the Voyager 2 encounter nine months later in 1981. Murray *et al.* (1997) called these strands F-α, F-β, F-γ, and F-δ. The strand F-γ is always present and may be identified with the "core" of the F ring. Showalter *et al.* (1992) described the F ring as a narrow core of centimeter-sized particles within a wider envelope of much smaller micron-sized dust. The larger particles sustain the ring and replenish material (a more detailed version is due to Barbara and Esposito, 2002). The Voyager view that the ring itself may have a limited lifetime and multiple reincarnations is confirmed and extended by Cassini results. The F ring, so evident in photos, is often not seen at radio wavelengths. The Cassini radio occultation, which is sensitive to particles larger than a centimeter, shows a core about 1 km wide, present about one third of the time. Cassini also saw a second, sibling core form (Albers *et al.*, 2012).

The known satellites of Saturn provide no obvious mechanism to produce this structure, but small undiscovered satellites around 5 km in radius could separate the observed strands (Murray *et al.*, 1997), which have nearly aligned orbits. Murray and his students have now extended this view from Cassini images (e.g., Atree *et al.*, 2012). Collisions and gravity both shape the ring. Regular structure arises from the gravity of Prometheus and Pandora (Figure 11.1). They form channels in the ring and also trigger the formulation of clumps. Moonlets about 10 km across strike objects in the F-ring core to produce jets of material, evolving into spirals. Smaller objects down to 10 m dimensions (Meinke *et al.*, 2012) can explain some fine structure and provide evidence for recreating the F ring (Esposito *et al.*, 2012; Charnoz *et al.*, 2009).

Bosh *et al.* (2002) have detected a slight inclination of the F ring from the ring plane crossing observations of 1995, with $i = 0°.0067$. Cassini confirms this orbit, at least in the average sense. But, on any given day, the F ring may be 20–50 km away from its predicted location. We see no color dependence of the ring's

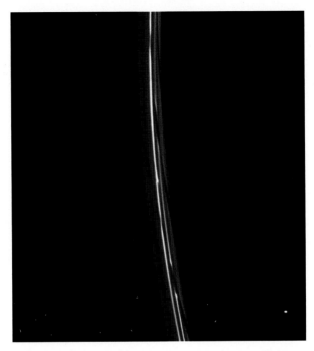

Figure 11.1 A train of diagonal channels in Saturn's F ring follows behind the moon Prometheus. Each of these features was created during a previous close approach of Prometheus to the ring. When the moon is at the furthest point in its orbit of the planet, it strays close to (and often into) the F ring. The resulting gravitational disturbance leaves behind the channels seen here. Atlas (30 km, or 19 miles across at its widest point) appears at lower right. This view looks toward the unilluminated side of the rings from about 27° above the ring plane. The image was taken in visible light with the Cassini spacecraft narrow-angle camera on July 5, 2008. The view was acquired at a distance of approximately 1.1 million km (675 000 miles) from Saturn and at a Sun–Saturn–spacecraft, or phase, angle of 34°. Image scale is 6 km (4 miles) per pixel. (PIA10448; courtesy NASA/JPL/Space Science Institute.)

total brightness, indicating that the particles are mostly large compared with the wavelengths of visible light, with size $r \geq 10$ μm. This contradicts the steep size distribution and abundance of dust inferred by Showalter *et al.* (1992). Bosh *et al.* (2002) speculate that this difference may be due to temporal or longitudinal variability. French *et al.* (2012) provide spectacular confirmation of the time variation of the F ring by comparing the Voyager and Cassini images.

11.2 F-ring shepherds

The narrow rings of Uranus posed a problem: what maintains them separate, unlike Saturn's broad rings? Goldreich and Tremaine (1979) proposed that small (then

Figure 11.2 A quartet of small satellites is nestled just outside of Saturn's classical ring system in this January 31, 2002, Hubble image. The scattered light from the bright rings has been removed by differencing successive frames taken in the same filter for the region outside of the A ring, revealing Janus, Epimetheus, Prometheus, and Pandora in isolation. The brightness of these moons has been increased by a factor of 30 to render them more visible. The F ring, not visible in this rendering, lies between Prometheus and Pandora. (From French *et al.* (2003); reproduced with permission from Elsevier.)

unseen) satellites confine them (see Chapter 5). A shepherd satellite on either side holds the ring in place. The Voyager encounters led to the discovery of a host of new satellites around Saturn. Two of these, Prometheus and Pandora, are located astride the F ring, apparently confirming Goldreich and Tremaine's model (see Figure 11.2). Unfortunately, we find that the F ring is closer to Prometheus, the larger inner satellite, and further from Pandora, the smaller satellite outside the

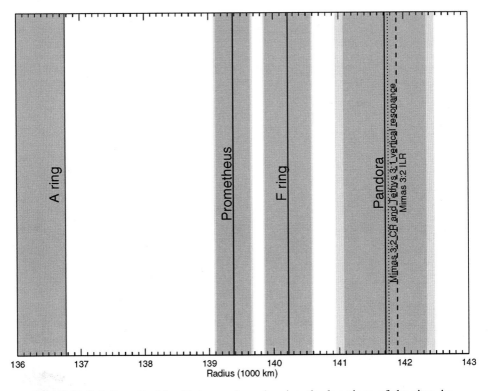

Figure 11.3 A "map" of the F-ring region, showing the locations of the ring, its neighboring satellites Prometheus and Pandora, and the locations of the important corotation and inner Lindblad resonances. The dark shading on either side of the F ring, Prometheus, and Pandora represents the radial range *ae* that these objects occupy during the course of their eccentric orbits. Semi-major axis *a* and eccentricity *e* for the satellites are from French *et al.* (2003) while *a* and *e* for the F ring are given in Bosh *et al.* (2002). The lighter shading shows the extent of *ae* when the eccentricity is augmented by 3σ, where σ is the formal error of the fit. (From French *et al.* (2003); reproduced with permission from Elsevier.)

ring (see Figure 11.3). In an evolved system, we would expect all the torques to be in balance between the shepherds and the rings. For the F-ring system, this would mean that Pandora should be closer, so that its shepherding influence on the ring is just the same as that of Prometheus. Perhaps the F ring is young and has not achieved this balance yet; see Chapter 10. Alternatively, the perturbations from these shepherds are significantly stirring up and agitating the rings. Perhaps the F ring persists in some "safe haven" from these destructive influences. This could explain its particular location, in the opinion of Jeff Cuzzi.

This puzzle became more complicated when these shepherds were seen again by the Hubble Space Telescope during the period in 1995 when Saturn's ring plane was aligned with the Earth. At such ring plane crossings, the edge-on rings are much dimmer and the smaller satellites near the rings become visible. Dick French and his colleagues have an observing program with Hubble that has followed these satellites several times a year since then, despite the glare from Saturn's rings. The surprising result from the ring plane crossing observations was that Prometheus and Pandora were not at their predicted locations (French *et al.*, 2003); although between 1994 and 2000, the motions of the two satellites were simple precessing elliptical orbits. Pandora also shows a 585-day libration with a nearby $3:2$ resonance with Mimas. Between December 2000 and September 2001, both moons abruptly changed their orbits by roughly equal and opposite amounts, suggesting a common cause for their orbit changes (see Figure 11.4).

Thus, although the moons were originally hailed as confirming a simple model of ring confinement, their positions and motions cast doubt on this simple explanation. Both shepherds moved *toward* the F ring. The shepherding moons would be expected to move *away* (although 1000 times more slowly). The longitudes seen from 1994 to 2000 are explained by Prometheus being 0.31 km closer to the F ring and Pandora 0.20 km closer, compared with their orbits observed by Voyager in 1980. Since 2000, each has moved still closer, Prometheus by 0.33 km and Pandora by 0.42 km. These possibly coordinated orbital changes argue for a direct exchange of energy between the two moons.

A related problem is the question of the age of Saturn's A ring. Jack Lissauer (1985) calculated that the exchange of angular momentum due to density waves raised in the A ring would push the moon Prometheus away from the ring, reaching its present location after just 10^7 years. The evolution is so fast that it requires either that the A ring be very young ($\lesssim 10^7$ years), or that the momentum be transferred to other, larger moons, which would evolve more slowly. François Poulet and Bruno Sicardy (2001) investigated the long-term behavior of the Prometheus–Pandora system and explored the mechanisms that might slow down this outward evolution. Unfortunately, capture is inefficient and short-lived. Chaotic motion destroys the resonances and probably also the satellites through collisions. If the moons are destroyed, perhaps new ones could be created at the ring edge, as suggested by Sebastien Charnoz.

Borderies *et al.* (1984) noted that the satellite orbital evolution could be acceptably slow if Pandora and Mimas were in a $3:2$ resonance. The effects of the nearness of this resonance (a 585-day libration) are clearly seen in the data analysis of French *et al.* (2003). Mimas, which is much larger, can absorb the excess angular momentum transferred from the rings and the shepherds and then pass it on to Tethys by virtue of their $4:2$ resonance. Sadly, Pandora's orbit is 49 km inside the Mimas $3:2$ corotation resonance, so this explanation does not work.

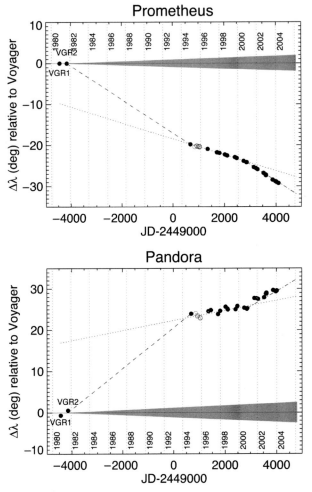

Figure 11.4 The puzzling long-term behavior of Prometheus and Pandora becomes especially evident when their orbital longitudes relative to the Voyager predictions are shown over the full time interval since the Voyager encounters. The relatively linear trends seen during the 1994–2002 Hubble Space Telescope observations (dotted lines), extrapolated back in time, do not match the observed positions of the two satellites at the times of Voyagers 1 and 2. The 3σ uncertainty in the predicted longitude at any given time, propagated from the Voyager observations, is shown as a shaded wedge in each panel. The mean motions of both satellites during 1980–2004 must have been different from their present values, in order to be compatible with Voyager observations. In 2001–2002, French *et al.* (2003) witnessed yet another change in the mean motions of the two satellites, as illustrated by the dot-dashed lines. (From French *et al.* (2003); reproduced with permission from Elsevier.)

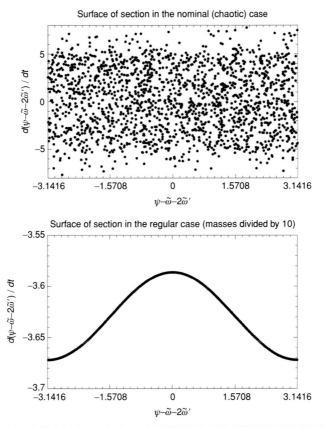

Figure 11.5 Plots of successive relative positions of Prometheus and Pandora calculated over a 3000-year interval show chaotic behavior (top). If the masses are reduced by 10 in the calculation, no irregular behavior is seen (bottom). This confirms that the gravitational interactions of the two shepherds leads to the chaotic behavior shown in Figure 11.4 (From Goldreich and Rappaport (2003); reprinted with permission from Elsevier.)

11.3 Explanations for the shepherd wandering

Two of the authors of that 1984 paper (Borderies *et al.*, 1984), Peter Goldreich and Nicole Rappaport (formerly Borderies), have completed new research attesting that the two shepherding moons are interacting chaotically. They can fully account for the observed wandering by gravitational interactions between the two satellites (Goldreich and Rappaport, 2003). This unexpected result was proposed by them almost 20 years earlier. Their numerical integrations (see Figure 11.5) confirm their earlier suggestion. They note that "for us, it is almost like a dream come true." The chaotic interactions would naturally explain the coordinated motions of the shepherds and the magnitude of their deviation from the locations predicted

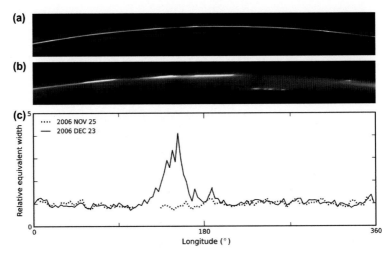

Figure 11.6 Cassini observations of Saturn's F ring show a bright feature suddenly appearing in late 2006. (a) A portion of the F ring (longitudes 146.3–155.4) as seen on November 25, 2006 (image N1543214356). (b) A similar portion (longitudes 148.1–157.1) as seen on December 23, 2006 (image N1545564196, corresponding to JPL Photojournal image PIA08863). (c) The longitudinal profile of the movies from observation IDs ISS_033RF_FMOVIE001_VIMS taken on November 25, 2006 (dotted) and ISS_036RF_FMOVIE001_VIMS (solid) taken on December 23, 2006. (French *et al.* (2012); reprinted with permission from Elsevier.)

from Voyager observations. The largest effects are seen when the slightly elliptical orbits of the two moons are anti-aligned, so that they pass by each other when the inner shepherd is furthest from Saturn and the outer closest (or vice versa). Such anti-alignments of the orbital apsides occur every 6.25 years. This was the situation in late 2000 and it occurred again while the Cassini space mission orbited Saturn in early 2007. Naturally, the Cassini investigators were watching the shepherds closely at that time, but no exceptional events were seen.

11.4 Cassini sees F-ring brightening

Cassini images now show that the F ring is about twice as bright during the Cassini mission as during the Voyager flybys of 1980 and 1981. It is also three times as wide and overall contains more integrated scattering areas. At some times the ring is optically thick and opaque to sunlight and stars that pass behind the rings, which was not observed by Voyager. Despite the change since Voyager, this ring has stayed much the same during the Cassini mission. As also seen during the Voyager flybys, bright spots have appeared! Cassini saw a large bright feature in 2006 that initially quadrupled the brightness of the entire ring, disappearing with a half-life of 91 days. See Figure 11.6.

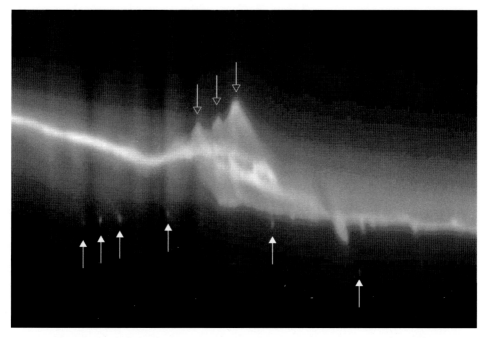

Figure 11.7 Reprojected Cassini image indicating evidence for small ($\lesssim$1-km) objects (filled arrows) in the F ring, along with possible corresponding features they create (open arrows). (Reprinted from figure 13.26, Colwell *et al.* (2009b), with kind permission from Springer Science and Business Media.)

Other images and movies show that the F ring continually changes. This dusty band of rubble contains a long-lived core and some strands that come and go. Murray's strands α–δ are no longer seen. The primary perturbations are from Prometheus, which carves channels into the ring at each passage (see Figure 11.1). In addition, the sudden brightenings caused by dust released in collisions shear out into spirals and jets (Figure 11.7). A particularly bright strand created in late 2007 resembles the main F-ring core, but 100 km away from it. Stellar occultations show opaque and partially transparent bodies ranging in size from 30 m to 1200 m. Because these bodies come and go with multiple lives, I have nicknamed these features detected by Cassini stellar occultations after cats, for example, "Mittens" and "Pywacket." These *kittens* may represent the larger bodies in the F ring, which are mostly unseen in the spacecraft images, until collisions release a cloud of dust.

This model is similar to the modern image of a circumstellar debris disk, such as that for Beta Pictoris (see, e.g., Lagage and Pantin, 1994), in which unseen small-bodies belts, stirred by planets, produce a dusty disk visible because of its infrared excess. Charnoz (2009) and Murray *et al.* (2008) proposed that the

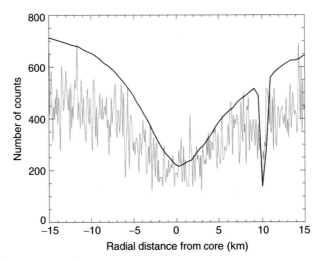

Figure 11.8 Simultaneous VIMS (solid, smooth curve) and UVIS (thin curve) occultation profiles showing the feature Pywacket approximately 10 km outside the F-ring core. (From Esposito *et al.* (2008); reprinted with permission from Elsevier.)

population of moonlets exterior to the core (among which the satellite designated S2004/S6 is a member) regularly collides with the population inside the core, releasing material whose orbital motion forms structures named spirals (Charnoz *et al.*, 2005) or jets. Several moonlets that could be members of this putative belt have been discovered in Cassini images (Porco *et al.*, 2005; Murray *et al.*, 2005, 2008) and in stellar occultations (Figure 11.8). The F ring is also famous for its shepherding moonlets, Pandora and Prometheus, which were believed initially to confine the ringlet radially (Goldreich and Tremaine, 1979) and are known to interact chaotically; see above (French *et al.*, 2003; Goldreich and Rappaport, 2003). However, the present understanding is much more complex, and it is not clear if this mechanism is really responsible for the F ring's narrowness.

Very-high-resolution images of the F-ring core (Murray *et al.*, 2008) reveal a wealth of kilometer-scale dynamical structures that remain to be explained. Conversely, the envelope and strands that surround the core are made of micrometer-sized dust, shining at high phase angles with a steep particle size distribution ($q \sim 4.6$) (Showalter *et al.*, 1992). Showalter (1998, 2004) proposed that transient bright features are dust clouds generated by meteoroid bombardment, whereas other authors suggested that the local collisional activity implying moonlets (or clumps) could be the cause of these events (Poulet *et al.*, 2000; Barbara and Esposito, 2002; Charnoz, 2009). Recent Cassini image data seem to support the latter model (Charnoz *et al.*, 2005; Charnoz, 2009; Murray *et al.*, 2008).

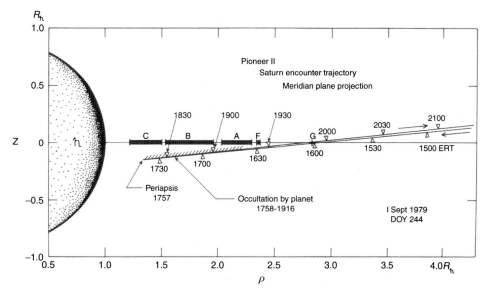

Figure 11.9 Geometry of the Pioneer 11 Saturn encounter, adapted from Van Allen *et al.* (1980), where ρ is the planetocentric distance, and z is the distance above or below the ring plane in Saturn radii. (From Cuzzi and Burns (1988); reproduced with permission from Elsevier.)

It seems that the F-ring origin is linked to the origin of a population of parent kilometer-sized moonlets. Cuzzi and Burns (1988) proposed that a small moon was destroyed in the past, whose fragments are slowly eroding today. Conversely, Barbara and Esposito (2002) suggested that there is ongoing accretion in the F-ring core, producing clumps and moonlets, whose subsequent collisional erosion produces the F ring.

11.5 Origin and history

The F ring's location outside the classical Roche limit (see Chapter 5), its brightness variations, and its multiple strands are challenging for any simple model of ring formation. In my view, the key is the interrelation of moons and rings – not just in the creating of the structure we see, but also in the origin and fate of the ring itself. Cuzzi and Burns (1988) explained the abrupt depletions in magnetospheric particles seen by Pioneer 11 experiments as due to small moonlets in a belt 2000 km wide surrounding the F ring. In 1979, Pioneer 11 made the closest approach to Saturn of any spacecraft, flying through the plane of Saturn's equator and below the rings (see Figure 11.9). This was some 16 hours after the F ring was discovered in grainy spin-scan images. The motion of charged particles in Saturn's Van Allen belts is quite complex (see Van Allen *et al.*, 1980, and Figure 11.10). Particles

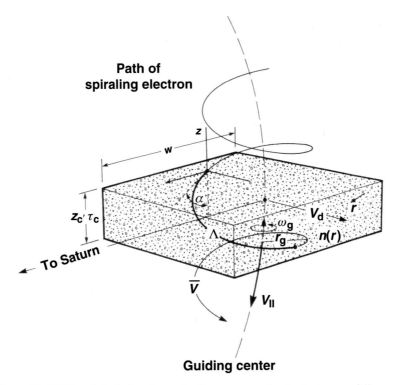

**Path of
spiraling electron**

Guiding center

Figure 11.10 Sketch depicting the path of a magnetospheric electron rapidly gyrating around magnetic field lines at gyrofrequency ω_g with gyroradius r_g and equatorial pitch angle α while latitudinally bounding between magnetic mirror points
at high latitude. At equator crossing, the electron's velocity is $\overline{V}$ and has a component $V_\parallel$ along the field line. The electrons undergo a slow longitudinal drift with
velocity V_d relative to an absorbing cloud of orbiting particles in the ring plane,
which is shown as having vertical thickness z_c, vertical optical depth τ_c, and radial
width w, and containing particles of radius r that occupy a volume with density
$n(r)$. Λ is the total path length of the electron on each vertical pass. (From Cuzzi
and Burns (1988); reproduced with permission from Elsevier.)

spiral around the magnetic field lines and bounce between the Saturnian polar
regions, while slowly drifting and diffusing. When the electrons and ions collide
with solid particles in a ring, they are absorbed, and a hole or "dropout" appears
in the data of magnetospheric experiments that cross that field line elsewhere in
the Saturn system. This absence of particles was observed on five occasions as
Pioneer 11 passed near the field lines that cross the F ring. Inconsistent with the
initial reports, these phenomena cannot be attributed to the shepherds Prometheus
and Pandora nor to a continuous ring (Van Allen, 1982, 1983). To explain this
phenomenon, Cuzzi and Burns proposed a belt of 0.1-km to 10-km radius objects.
When these objects collide with each other, they produce clouds of ejected material

that cause the observed magnetospheric depletions. This is strongly supported by Cassini observations; see Figures 11.6 and 11.7.

Every 15 years or so, the Earth and Sun cross Saturn's ring plane (see Figures 2.3 and 2.4). During the 1995/96 edge-on orientation ring, Earth-based observers had the opportunity to detect new objects near the outer edge of the main rings that are normally lost in the glare of light reflected by the rings. Surprisingly, numerous small bodies were reported. Within the errors of observation, these were within the F ring (Poulet *et al.*, 2000). However, their size, brightness, and in some cases disappearance, are inconsistent with actual solid bodies. Instead, they are more likely interpreted as temporary clouds of regolith resulting from collisions of parent bodies, as proposed by Cuzzi and Burns (1988).

Mark Showalter (1998) reanalyzed 1980 and 1981 Voyager images of Saturn's F ring with special attention to very transient bright features he called "burst" events. Showalter interpreted these phenomena as due to meteoroid bombardment of the F ring. Barbara and Esposito (2002) have criticized this explanation, preferring to explain them as due to collisions between unconsolidated parent bodies in the F-ring region. The disruption of these rubble piles provides a short-lived release of material that temporarily resembles a new satellite within the ring. This view is now generally accepted. French *et al.* (2012) compare the latest Cassini images. They show that the ring is now twice as bright as when seen by Voyager. Furthermore, it is also three times wider and more opaque too. Transient features continue to appear, as in 2006 when the ring became roughly two times brighter, a change that persisted for three months (Figure 11.6).

Barbara and Esposito's updated version of Cuzzi and Burns' (1988) model includes both accretion and fragmentation over a broad range of ring-particle sizes. The competition between these two processes results in a bi-modal size distribution with abundant dust as well as a significant number of larger bodies. More recent results by Meinke *et al.* (2012) cast doubt on their conclusion. The bi-modal distribution may result from the sharp size cutoff assumed by Barbara and Esposito. A similar phenomenon might explain "bumps" observed in the distribution of sizes for objects in the Kuiper Belt and asteroid belt, if the real distribution of larger objects has an actual sharp cutoff.

Far outside the Roche limit, we expect rapid accumulation of small bodies into a single satellite. Within the Roche limit, unconsolidated bodies would be disrupted by tides. Near the Roche limit, the physics is more complicated, and both processes occur. Simulations in this broad, "tidally modified" region show that accretion is only possible for two colliding bodies that differ substantially in mass (Canup and Esposito, 1995). By contrast, two bodies of similar size will exceed their *mutual Hill sphere*, which defines the moon's sphere of influence (Ohtsuki, 1993), and fail to become gravitationally bound. This feature leads naturally to a bi-modal distribution of ring particles. In such a distribution, the largest particles are unconsolidated aggregations of many smaller particles held together only by their self-gravity.

Barbara and Esposito (2002) agreed with Cuzzi and Burns, and concluded that the F ring itself may be only the latest recurrence of a relatively recent collision between the largest of these yet unseen objects. A more detailed proposal is from Salmon *et al.* (2010). Collisions between the ring-moons replenish the rings, and reaccretion replenishes the small moons (Poulet and Sicardy, 2001; Esposito *et al.*, 2008, 2012). How efficient and long-lasting this process of "cosmic recycling" can be remains an open question.

Bonnie Meinke studied 101 stellar occultations by Saturn's F ring. She found numerous small structures probably caused by elongated clumps of ring particles that we named after famous cats (living and dead). As I explained earlier, the idea is that these clumps are transient, coming and going as in the simulations by Mark Lewis and Glen Stewart, and thus possessing multiple lives! Saturn's F ring provides a natural laboratory for direct observation of accretion and disruption processes. Among the structures seen in its meager 10-km width are jets, strands and moonlets (see Figure 11.7). The nearby moons Prometheus and Pandora stir up ring material and create changing structures on timescales of days to decades. Meinke developed a numerical model that includes the stirring and its ability to trigger clumping. By increasing aggregation, she was able to match the distribution of "kittens" seen by Cassini (Meinke *et al.*, 2013). The wakes created by satellites form higher density regions; the boosted accretion helps create aggregates, whose size distribution matches the Cassini occultation results. Her conclusion is clear: accretion is manifestly overcoming fragmentation and disruption in the F ring.

As an example, the F ring now apparently contains substantially more material than it did 30 years ago. French *et al.* (2012) suggest that the F ring spawns it own impactors. Since the variations do not seem linked to the 17-year cycle of stirring by Prometheus, this may imply a delay while stirred up material reaccretes or, perhaps, just indicates the random stochastic nature of ring evolution and rejuvenation.

11.6 Summary

Saturn's F ring highlights the processes of origin, fragmentation, and accretion, and the short ring lifetimes that are found in both broad and narrow rings. It is a microcosm for accretion: close to the Roche limit, yet transparent enough to observe changes. The Cassini mission has made multiple, long-term observations of this mysterious ring but has not found any associated moonlet belt, only transient objects in and near the F-ring core. The spectacular changes underline that rings are even more dynamic than expected.

12

Uranus' rings and moons

Two separate teams of astronomers prepared to watch the star SAO 158687 as Uranus passed in front of it in 1977. The major objective was to study the planet's atmosphere, but the careful observers turned on their instruments early, and kept observing long after the planet passed. Both at the Perth Observatory in Australia and aboard the Kuiper Airborne Observatory flying over the Indian Ocean, they saw the star blink out several times. This was quickly interpreted as rings surrounding Uranus, as the occultation events were symmetric. The Kuiper team named the rings alpha, beta, gamma, delta and epsilon in order of increasing distance from Uranus; the Perth group named their discoveries 1–6. This explains the current nomenclature. (See Table 3.1.) New rings have followed the Greek nomenclature: λ, ν, μ, giving a total of 13 named rings, ranging from 38 000 to 98 000 km from Uranus' center.

12.1 Description

Most Uranian rings are dark, opaque, and only a few kilometers wide. The ε ring is the brightest and broadest of the narrow rings, with its width varying from 20 km to 96 km; see Figure 12.1. The rings 1986U2R (the designation is from the Voyager Uranus flyby), λ, μ and ν are dusty, the other rings are composed mostly of macroscopic particles up to radii of about 10 μm. The discovery by Hubble of two more rings along with associated moons was reported by Mark Showalter and Jack Lissauer: they describe their findings as a "second" ring–moon system around Uranus; see Showalter and Lissauer (2006). Imke de Pater and others (2007) observed with the Keck telescope the moment when the Earth passed through the ring plane. This equinox observation allowed observers a long path of sight through the rings, which made the optically thinner rings much brighter. In fact, the ζ ring was the brightest at that moment. The fact that the observing geometry of Voyager was distinctly different from Keck (see Chapter 15) means that each

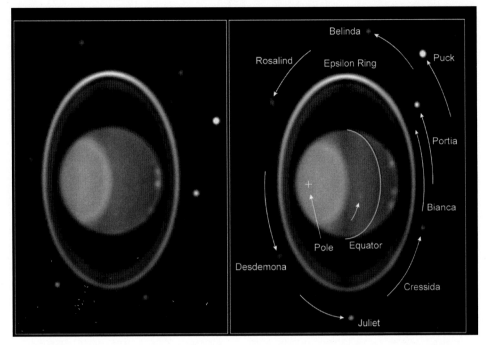

Figure 12.1 Uranus' clouds and rings are visible in this Hubble Space Telescope image, from July 28, 1997. The wavelengths for "blue," "green," and "red" are 1.1 μm, 1.6 μm and 1.9 μm, respectively. The rings of Uranus are quite prominent in the near infrared. The widest and brightest part of the E ring (Uranus' brightest ring) is at the top of the image. Fainter rings are visible inside it. Space Telescope Science Institute. (PIA 01278; courtesy of NASA/JPL/STSI.) For color version, see Plates section.

observation was sensitive to different particle sizes (see Figures 12.2 and 12.3). Alternatively, the distribution of dust may have changed, or perhaps aggregates may have disaggregated. These changes could be explained by dynamical interactions between clumps in the rings, which shed clouds of dust after collisions. This would be similar to the varying behavior of Saturn's F ring reported by French *et al.* (2012), which has also shown major differences from the earlier Voyager observations. See Chapter 11.

12.2 Dynamics and origin

The Uranian rings were the first narrow rings to be discovered. Their narrowness and sharp edges highlighted the issue of confinement (see Section 5.2), which led Goldreich and Tremaine to propose shepherding moons, one each outside and inside the ring, to restrain its particles. The moons gravitationally interact with the ring-particle system, transforming momentum to limit the ring spreading. They thus act as donors and sinks of angular momentum, but through conservation of

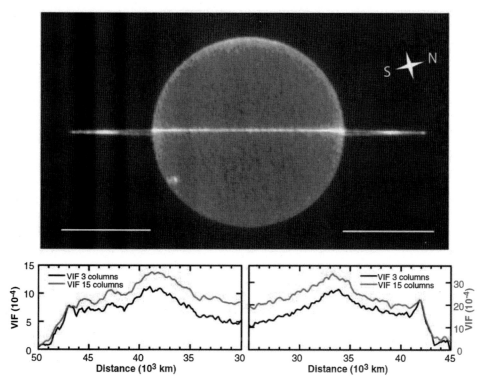

Figure 12.2 The dark side of the rings of Uranus, as imaged by Keck adaptive optics. The image is rotated ~90°, so the ring plane is oriented horizontally, and celestial north points toward the right (as indicated by the compass rose). Below the figure are profiles of ring intensity versus projected radial distance. Horizontal white bars in the image indicate the approximate radial extent of these profiles. Two versions of each profile are shown: one integrated over three pixels normal to the ring plane (i.e., three columns in the image before rotation; black line) and the other over 15 pixels (red line). The broader integral captures more of the rings' light but shows somewhat less detail. The y-axis is "VIF" or vertically integrated I/F. Here, I/F is a dimensionless quantity, where I is intensity and πF is the solar flux density. Note that the scales for VIF differ for the two integrals, with the black profile's scale shown at left and the red profile's scale shown at right. (From de Pater *et al.* (2007); reprinted with permission from *Science*.) For color version, see Plates section.

angular momentum, they spread apart themselves. The fact that no shepherds have been found other than Cordelia and Ophelia means this theory still provides an incomplete explanation of ring confinement. Similarly, no shepherds have been found for the narrow ringlets in Saturn's Cassini Division, a continuing puzzle to Cassini observers.

The dust in the rings is naturally created by collisions between the largest particles and by meteoritic bombardment. Because the small particles are subject to many non-gravitational forces, this dust has a short lifetime (100–1000

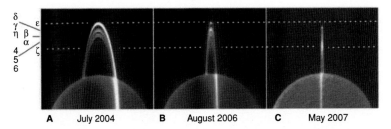

δ
γ ε
η β
      α
4
5   ζ
6

A   July 2004   B   August 2006   C   May 2007

Figure 12.3 Comparison of the lit and unlit sides of the rings of Uranus. (A) The lit side in early July 2004, when the angle B to Earth was 11° and the ring opening angle B_o to the Sun was 13.2° (5). (B) The lit side on August 1, 2006 when $B = 3.6°$ and $B_o = 5.2°$. (C) The unlit side on May 28, 2007 when $B = 0.7°$ and $B_o = 2.0°$. The dotted lines show the position of rings ε (top line) and ζ (bottom line). The pericenter of ε was near the tip of the ring in 2006, was at about the eleven o'clock position in 2004, and was at about the two o'clock position in 2007. (From de Pater *et al.* (2007); reprinted with permission from *Science*.) For color version, see Plates section.

years) as is similar for the Jupiter and Neptune rings, and the ethereal rings of Saturn.

The source bodies that are parent moonlets and large ring particles are typically too small to be seen by Voyager, although, in some other rings, Galileo, New Horizons, and Cassini have detected them.

12.3 Summary

Overall, the Uranian rings (and Neptune's as well, although they are further from the planet; see Chapter 13) provide an intermediate case between the faint, dusty Jovian rings and Saturn's broad, complex rings. They are more massive and higher in optical depth than Jupiter's. Their source bodies are not as obvious. They are composed of darker silicate material like Jupiter's rings. They must be younger than Saturn's broad rings and thus continually replenished. However, Saturn's narrow, dusty ringlets embedded in the Encke Gap provide a reasonable comparison to the narrow Uranian rings. Jupiter's rings resemble the Uranian dust bands. Like other ring systems, the Uranus rings are more complicated and dynamic than originally expected. They contain unexplained features. They are younger than the planet they encircle, implying continuing creation of dust and renewal or recycling over the age of the solar system.

13

Neptune's partial rings

Mark Showalter of the SETI Institute wrote a "Perspective" in *Nature* in 1999 titled "Neptune's misbehaving rings." He noted in this piece that, as is often true for planetary rings, Neptune's rings did not behave as we predicted! With recent progress in the past years, our understanding has now advanced: this allows me to soften his title. Nonetheless, our understanding is still incomplete. New space missions to Neptune may be needed to resolve the uncertainties and check the latest predictions.

Showalter placed the latest difficulty with Neptune's rings in the context of the long history of scientists oversimplifying our picture of planetary rings. Naturally, at first we had a picture of Saturn's rings (the only known ring system before 1977) as broad, featureless, flat, equatorial, circular, and symmetric. This picture was consistent both with the observations and the expectation that collisions between the particles in the ring would quickly smooth out inhomogeneities. We now know that all the giant planets have ring systems, and each simplification has been contradicted by more recent and more detailed observations. At the same time, the simple theoretical expectations also need to advance so that we can understand how rings are *not* violating the laws of physics. Neptune's rings provide an extreme case, one that is still incompletely explained.

The Neptunian rings contain a large fraction of micrometer-sized dust, 20–70%, resembling Jupiter's rings, and significantly different from the main rings of Saturn and the narrow rings of Uranus (see Chapters 3, 12). The ring particles are dark, probably made with ice contaminated with organics. Their red color and *albedo* resemble the Uranian ring particles and the inner moons of Neptune. Like Jupiter's rings, they include faint dusty rings and even fainter broad and dusty rings.

The same technique of stellar occultation that led to the discovery of the Uranus rings in 1977 provided the first information about Neptune's ring system. In the

1980s, ground-based telescopes observed a number of bright stars pass behind Neptune. Initially, the stellar occultations showed no opaque features, or occasionally a blockage on only one side of Neptune. Since a one-sided ring was considered impossible, these rare events were interpreted as experimental errors. Members of the American Astronomical Society's Division for Planetary Sciences, including a famous non-professional, predicted rings around Neptune on general principles. At a subsequent professional meeting, James Elliot, the discoverer of the Uranian rings, explained his reasons why rings did not exist around Neptune. An indisputable occultation in 1984 showed the rings to be partial – two separate observers saw occultations on only one side of Neptune. This was a completely unexpected result. In looking at all the occultations from the perspective of partial rings, the rings appeared to be present only about 10% of the time. Nothing in the collective experience of astronomers would have predicted this phenomenon, but, within a year, two models were developed to explain this, one by Jack Lissauer (1985) and another by Peter Goldreich *et al.* (1986). We now believe that aspects of both models are necessary to fully explain the Neptune ring system.

13.1 Dynamical explanations

The dynamical models for partial rings draw on two important ideas from the modern era: unseen small moons and satellite resonances. Lissauer (1985) suggested that an embedded moonlet would cause ring particles to be trapped like Jupiter's *Trojan asteroids* in stable positions 60° away from the moonlet. The ring particles would librate about the exact longitude (move in angle relative to this position, just as the libration of the Moon causes a little more than half its surface to be visible). Recalling the model of the pendulum developed in Chapter 6, we can see this oscillation as similar to pendulum motion about equilibrium. As seen in the frame rotating with the moonlet, the total of all the ring particles sharing this motion would smear out into two long arcs preceding and trailing the embedded moonlet. One difficulty with this model is that the Trojan points are actually maxima of the potential energy. As ring particles collide, they lose energy in the inelastic collisions and the particles drift away. If they drift too far, they are lost from the Trojan point and then circulate in a more distant or a closer orbit relative to the planet, so they regularly pass by the moonlet instead of being trapped in a small range of longitude. Eventually, all of the particles would soon be lost, diffusing away like gas molecules into a vacuum. To resupply the energy, Lissauer proposed another moon interior to the ring arcs whose resonance (perhaps an outer Lindblad resonance; see Chapter 6) could provide a source of energy and momentum transfer for the ring particles. The continuing pumping would allow the partial rings to last a long time.

13.2 Voyager observations

Voyager 2 flew by Neptune in August 1989, providing the first and only close-up views of Neptune's atmosphere, satellites, and mysterious rings. The pictures showed a faint ring, invisible from Earth, with a few brighter arcs embedded within it. Andre Brahic of the Paris Observatory said at a press conference that this resembled an irregular type of sausage produced in the south of France, whose links are variable in length and width. All the arcs were within a sector of 40° length. Frustratingly, these initial observations appeared inconsistent with both proposed dynamical models. Not surprisingly, based on its previous experience at Jupiter, Saturn, and Uranus, Voyager also discovered new moons circling Neptune, including Galatea (about 75 km in radius), which orbits 980 km inside the ring containing the arcs (now known as the Adams ring, after one of the mathematicians who predicted the planet Neptune's existence in 1842 from study of its gravitational effects on Uranus). At least three other rings were discovered by Voyager (see Table 3.1). The three major arcs were named Liberté, Egalité, and Fraternité. A fourth arc, named Courage, was identified later.

Voyager observed the bright star σ Sgr as it passed behind the rings (Figure 13.1), allowing exact determination of the ring location and albedo. The rings are very dark, reflecting only about 5% of the incident sunlight. The arcs are dustier than the optically thinner parts of the continuous ring that contains them. Colwell and Esposito (1990b) explained this as due to collisions among ring particles energetic enough to knock dust off their surfaces. Since this dust is rapidly lost (radiation effects remove 1-μm particles in about a century), the arcs and the dustiness both require active processes to maintain them, and thus indicate young rings. Just *how* young is an open question (see below). An attempt to observe a radio occultation by allowing the radio beam from the Voyager spacecraft to pass through the rings on its route to Earth failed, because it unluckily did not intersect any of the arcs. Voyager pictures clearly showed some small clumps of material within individual arcs (see Figure 13.2).

13.3 Porco's model

In 1991, Carolyn Porco (then at the University of Arizona, now at the Space Science Institute in Boulder, Colorado) showed that the Goldreich model could explain the arcs in the case that only some of the corotation sites are occupied. If the Adams ring is at the 42:43 *corotation inclination resonance (CIR)*, the combination of the Galatea nodal *regression* and the mean orbital motions of Galatea and a ring particle provide 86 stable libration sites, each 4.2° long (= 360°/86). Particles in each of these sites would behave like Trojan asteroids, librating about the exact resonance longitude, but with a total longitudinal excursion of 4.2° or less. To

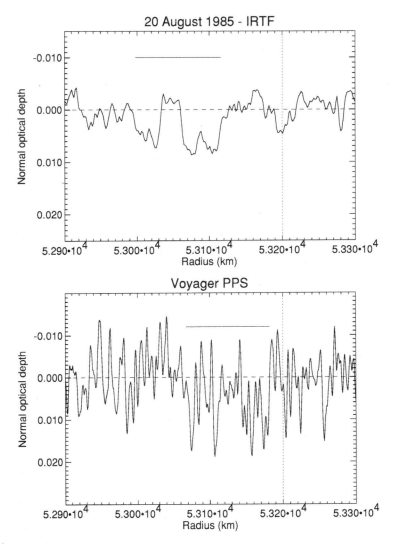

Figure 13.1 A comparison of optical depth profiles for the LeVerrier ring obtained from the August 20, 1985 and Voyager PPS occultations, on a common radial scale. The ground-based and PPS data have been averaged to radial resolutions of 8 km and 5 km, respectively. The overall widths and average optical depths of the two profiles are similar, although both are close to the noise level in their respective data sets. There is some evidence in both profiles for a two-component structure. The image-derived mean radius for this ring of $53\,200 \pm 20$ km is indicated by the dotted line. The systematic astrometric uncertainty in the ground-based data is $\sim$35 km, while that in the PPS data is $\sim$5 km. (From Nicholson *et al.* (1995); reprinted with permission from Elsevier.)

Figure 13.2 A Voyager image of the two outermost Neptunian rings. Orbital motion is clockwise. The direction of motion of the spacecraft during the exposure, indicated by the trailed stars, has smeared the arcs to look much wider than their natural widths of ~15 km. (Courtesy NASA/JPL-Caltech.)

explain the observations that only several arcs are seen, it is argued that by chance only a subset of these 86 sites would actually hold significant amounts of material. If adjacent sites happen to be filled, the total arc length is a multiple of 4.2°. In this way, the Voyager observations can be compatible with the Goldreich model (see Figure 13.3).

Philip Nicholson (Nicholson *et al.*, 1990) of Cornell University analyzed all of the occultations by Neptune's arcs from ground and space and found two possible solutions for the arcs' mean motion (and thus, from Kepler's law, their average distance from Neptune). The first of these solutions fits within small errors of Porco's prediction. This was taken as a confirmation of the model, because otherwise it would be extremely coincidental for the observed ring and its arcs to fall exactly as predicted by theory. Nicholson explained that this solution determines exactly which arc (Liberté, Egalité, or Fraternité) was detected by every previous occultation (ground- and space-based) and shows that their longitudinal structure has persisted for a decade or more. When later it was discovered by Earth-based observations that the second of Nicholson's solutions was actually the correct one

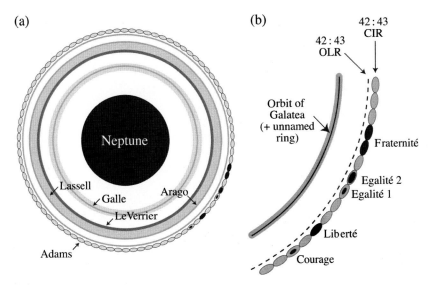

Figure 13.3 A schematic diagram of the rings of Neptune as viewed from the south pole showing (a) the location and names of the main rings and (b) the theory proposed by Porco (1991) to explain the existence of arcs in the Adams ring as the filled centers of libration of the Galatea 42 : 43 resonance. CIR, corotation inclination resonance; OLR, outer Lindblad resonance. (Adapted from Murray and Dermott (1999); reprinted with the permission of Cambridge University Press.)

(see Section 13.6), the entire theory and these identifications were called into question.

13.4 Problems with Porco's model

The CIR explanation was almost universally accepted, although a number of problems were evident (see, e.g., Esposito, 1993). The observed width of the ring was too large to fit the very narrow resonance condition ($\Delta r \leq 0.2$ km). Further, collisions among particles would remove the ring particles from the resonance, increasing the population of the diffuse background ring. Voyager saw many features smaller than a resonance width that were not explained by the model. Foryta and Sicardy (1996) showed that chaotic diffusion allowed particles to migrate from one resonant site to an adjacent one in an unpredictable manner. Esposito *et al.* (1997) calculated that all 86 resonant sites would be filled in about 10 000 years if this chaotic behavior were unchecked.

A number of authors (Lissauer and Sicardy, 1990; Esposito *et al.*, 1997) suggested that a small number of unseen moons could assist in arc confinement. Since the size of these hypothetical moonlets would range from 1 km to 20 km, they would be too small to be seen by Voyager or by the best Earth-based telescopes.

Unfortunately, any model hypothesizing moonlets has numerous free parameters, leaving the explanation underdefined and untestable. A general review of the Neptune rings by Porco *et al.* (1995) concluded that some combination of the models of resonance and embedded moonlets together could explain the origin and persistence of Neptune's partial rings. The embedded moons would be the natural result of a disruptive event where a moon was destroyed to create the rings, as proposed by Colwell and Esposito (1992, 1993). The rings of Neptune are interesting because they provide an intermediate case of the balance between destruction and accretion.

13.5 Salo's model

The most developed model of this sort is by Heikki Salo and Jyrki Hänninen (1998) at the University of Oulu, Finland. They carried out detailed numerical simulations, including collisions and the gravitational attraction among ring particles. They noted that the dust in the rings must be rapidly replenished and that impacts between the larger particles destroy the resonance locking. A change in a particle's semi-major axis of only 0.3 km removes it from the corotation resonance. This corresponds to a collision velocity of only $3 \, \mathrm{cm \, s^{-1}}$. Even less energetic collisions will cause a gradual change in location and thus eventual loss. Once the particles escape the resonance that maintains an arc, they continue to orbit Neptune in the continuous diffuse ring, which has a width of 50 km or so (this is much broader than the arcs themselves, which are only 10–15 km across). However, dust can be recaptured by recolliding with the larger bodies in the arcs. A major requirement of any model is that it explain the relative abundance of ring particles within the resonance locations compared with outside of it. Voyager observations give a ratio of about 30 : 1.

Salo's model followed that of Colwell and Esposito by producing dust when larger particles collide. These ring particles successively release dust in collisions and collect it between collisions. The ring particles are assumed to have a broad size distribution, with many small particles and fewer large particles, ranging in size up to several kilometers. Such a size distribution is the natural result of destruction of a small moon, which probably provided the initial birth of the Neptune arcs. This explains why the arcs are all concentrated on one side of Neptune: the detritus from the disruption of a small moon was captured into just a few adjacent corotation sites. It is very unlikely that this arrangement would happen by chance, a probability of less than 10^{-8} (Esposito and Colwell, 1992).

Salo and Hänninen (1998) proposed that several of the larger particles are trapped in each resonance site on orbits that are stable and never collide. These would be the largest fragments remaining of the original moons. Each is unfortunately too small to be seen by Voyager, or from Earth. Particles and dust escape the resonance but are confined between the largest particles: they execute horseshoe orbits, as the gravity of the larger particles in the next resonance reverses their direction as they approach it. This is similar to the case of the co-orbital satellites Janus and Epimetheus of Saturn. The visible arcs are thus *between* the exact corotation resonance longitudes – a significant difference from the Porco model, which imagines that we see the arcs *in* the resonance sites. Unfortunately, the existing observations are not accurate enough to decide between these two opposite predictions.

This model also gives a natural explanation of the smaller, unresolved clumps seen within the arcs: they are caused by collisions among the smaller particles contained in the arcs. Because an arc would occur between two large ring fragments (big enough to reverse the smaller particle motion), not every corotation site need have one fragment, and arcs could extend over several sites, allowing them to be longer than 4.2°, as observed.

The natural history of this proposed arrangement is straightforward. A small moon is destroyed. Its fragments are caught in corotation resonance with Galatea. The orbits of the largest particles evolve so that they are non-colliding; the smaller particles collide with each other and knock dust off their surfaces. The dust and small particles are held between the biggest fragments, covering one or more multiples of 4.2° in longitude. Although the smallest particles have only a fraction of the mass and are pushed about by their big brothers, they nonetheless dominate by number and also surface area: these are what we see in the Voyager pictures.

The evolution of non-interacting orbits in longitude is analogous to the Cordelia connection (see Chapter 10) in the Uranian rings. Objects that do not exist in stable resonance can experience destructive collisions or ejection from the system. A similar example may be seen in Pluto's 3 : 2 resonance with Neptune, which assures that the two planets never come close on their orbits. The Neptune ring arc moonlets are survivors, too.

13.6 The deadly 1998 observations

When Neptune's partial ring arcs were first observed by Voyager in 1989, it was believed they would be invisible from Earth. Instead, the pace of astronomical development was such that, within a decade, they were seen both from the ground using the technique of adaptive optics and by the Hubble Space Telescope. These observations sounded the death knell for the model proposed by Goldreich: that the arcs lay at the 42 : 43 CIR with Neptune's moon Galatea. The later models of Porco and Salo were also based on this hypothesis and thus in doubt. The publication of these results in *Nature* magazine in 1999 (Dumas *et al.*, 1999; Sicardy *et al.*, 1999) was the occasion for Showalter's article on "misbehaving" rings.

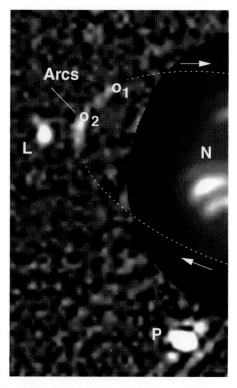

Figure 13.4 Initial image of Neptune's ring arcs obtained by the Hubble Space Telescope and its near-infrared camera NICMOS on June 3, 1998. Neptune's disk was positioned partly outside the field of view of the camera to reduce the scattered light coming from the planet. In this image, Neptune's contribution to the background has been modeled and removed. An image of Neptune's clouds has been superimposed for better clarity of the system geometry. The two circles (labeled 1 and 2) correspond to the positions of the middle-point of the trailing Egalité arc derived from the two possible solutions for the arcs' mean motion: $n_1 = 820°.1194\,\mathrm{d}^{-1}$ and $n_2 = 820°.1118\,\mathrm{d}^{-1}$. The letters L, P, and N mark respectively the positions of Larissa, Proteus, and Neptune, and the arrow shows the direction of motion of the ring arcs along their orbit. (From Dumas *et al.*, 1999); reprinted by permission from *Nature*, **400**, 733–5, © 1999 Macmillan Publishers Ltd.)

The two teams of observers were led by Bruno Sicardy of the Paris Observatory and Christophe Dumas of the Jet Propulsion Laboratory in Pasadena, California. Each was able to see the arcs (see Figures 13.4, 13.5, and 13.6) and several of Neptune's small satellites that had been discovered by Voyager in 1989. Each satellite was at a longitude in its orbit consistent with the Voyager predictions, but the arcs were not. They were out of place by about 20° with respect to the prediction from Goldreich's model, and in fact were quite consistent (within about

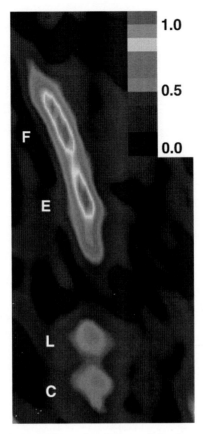

Figure 13.5 High-resolution false-color image of the four ring arcs of Neptune obtained by the Hubble Space Telescope's NICMOS near-infrared camera at a wavelength of 1.87 μm. A total of 22 dithered images of 208 s each were recorded on June 20 and October 22, 1998. A model of the planet's scattered light was subtracted from the individual images in order to (1) recover a background level close to zero at the distance of the arcs from Neptune's center, and (2) eliminate stray light produced by the secondary mirror support structure. The signal produced by the brightest arcs (Egalité/Fraternité) was at the noise level in the individual images. The arcs were extracted from the individual images by taking into account their motion in the reference system linked to Neptune. During the co-registration process, the data were resampled by a factor of 2 onto a finer grid by cubic convolution interpolation. The physical center point of the system of satellites and arcs was determined by using the inner Neptunian satellites (Proteus, Larissa, Despina, and Galatea) present in the images to fit their individual trajectory whose pole solution has been known since Voyager. The benefit of the recentering, shifting, and adding processes was to increase the resolution of those images to a point where the error of our measurement of the position of the arcs along the Adams ring was reduced to ±1.5°. The letters C, L, E, and F indicate, from leading to trailing, the respective locations of the arcs Courage, Liberté, Egalité, and Fraternité. (From Dumas *et al.* (1999); reprinted by permission from *Nature*, **400**, 733–5, © 1999 Macmillan Publishers Ltd.)

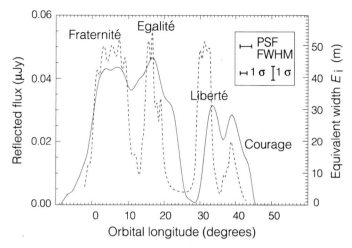

Figure 13.6 Brightness profile of the arcs along the azimuthal direction. The azimuthal profile of the arcs derived from Figure 13.5 is shown here (solid line) both as the NICMOS radially integrated flux in µ Jy per azimuthal degree and as the equivalent width (E_i) in meters. Relative brightness of the arcs from Voyager (1989) observations, normalized at Egalité, has been overplotted (dashed line) for comparison. Within the measurement uncertainties, the same overall structure was seen in both low and high phase angle Voyager images. The 1σ uncertainties in both the NICMOS observed flux and azimuthal position are also indicated as well as the width (full-width at half-maximum, FWHM) of the NICMOS point-spread function (PSF) at 1.87 µm. The most apparent changes in the *ring arcs'* morphology since 1989 are the widening of Egalité and the decrease in amplitude of Liberté. An effect induced by different scattering properties between high and low phase angles is not likely but cannot be entirely ruled out. (From Dumas *et al.* (1999); reprinted by permission from *Nature*, **400**, 733–5, © 1999 Macmillan Publishers Ltd.)

1° of longitude) with the second possible prediction from Nicholson, which all had ignored since it contradicted the theory. Thus, the rings were not misbehaving, but the 1998 observations showed the theory to be wrong. The corotation resonance model could no longer explain the confinement of the ring arcs. Without such confinement, they would smear out into a continuous ring in a matter of months, yet these latest observations showed they had clearly persisted for more than 15 years. It was this quandary that called the stability of the arcs into question. Dynamicists had to go back to square one as the possibility of more moonlets playing a role was highlighted again.

In my own case, puzzlement reigned. I put Neptune ring studies on the back burner. Although models by Salo and others had seemed to provide a frame-work for future progress, now everything was in question. I decided to concen-trate on Saturn's F ring, which shared a number of puzzling characteristics with

Neptune's rings, and unlike them would soon be visited close up by Cassini (see Chapter 11).

13.7 Namouni and Porco's alternative solution

When the Hubble and adaptive optics results in 1998 showed the arcs to be displaced from the CIR, their stability against the natural tendency of collisions to spread out the ring particles was once again unexplained. In 2002, Fathi Namouni and Carolyn Porco, both then of the Southwest Research Institute, discovered a neglected resonance that could explain the angular confinement of Neptune's ring arcs.

The former explanation that Galatea produces a set of equilibrium points rotating at a constant rate (the pattern speed; see Eq. (6.1)) is a result of combining its mean motion and vertical frequency (equivalently, its nodal regression) so that particles in the Adams ring see a periodic gravitational tug. If the ring particles' mean motion is near enough to the forcing, the particles are driven to librate about the stable equilibrium points. Unfortunately, as stated above, these equilibria are potential energy maxima, and collisions tend to remove the particles rapidly. In the previous explanation, Galatea counterbalances this by forcing the ring particles' eccentricity: this is an energy source. Despite the beauty and economy of this explanation, the 1998 observations show that the rings do not lie at the $86:84$ CIR as proposed. Fine-tuning of the system's parameters cannot shift the resonances the required 300 m outward. Adjusting the parameters would predict a systematic shift in the other Neptunian satellites discovered by Voyager, but the 1998 observation confirmed that these small satellites were at their predicted positions. Making the rings or Galatea much more massive could also work to shift the resonance back to the observed arc location, but the changes required are too large to be considered likely.

On the other hand, Namouni and Porco (2002) noticed that other resonances are more susceptible to changes that are less contradictory to our current knowledge. For example, it is possible to move the nearby *corotation eccentricity resonance* (CER $43:42$, where the resonance angle is $43\lambda - 42\lambda_g - \omega_g$, or $m=43$, $k=1$, $p=0$; see Chapter 6) by accounting for the gravitational pull of the arcs on Galatea's eccentric orbit. Although Voyager did not measure Galatea's orbit well enough to determine if it is non-circular, the inaccuracy of the measurements allows an eccentricity $e_g \approx 2.5 \times 10^{-4}$. A possible solution of ring mass and eccentricity can be found for $e_g \approx 10^{-6}$ and a ring mass 1/500 of Galatea's mass. If the Neptune ring arcs were created by disruption of a small satellite formerly in resonance with Galatea (Colwell and Esposito, 1990a, b; see Chapter 10), the original parent moon having this mass would have been about 10 km in radius. This is comparable to the size of the small moon Pan, found in the Encke Gap of Saturn's outer A ring; thus this proposal is not at all unreasonable.

Confirming the model proposed by Namouni and Porco will probably require close-up views of the Neptune system from a future space mission. Such a mission is still beyond the horizon of NASA's most ambitious longest-term plans but could confirm this prediction and provide a direct inference of the Neptune ring arcs' mass, by measuring the small residual eccentricity of Galatea's orbit around Neptune.

13.8 Keck observes the Neptune rings with adaptive optics

The Voyager observations of 1989 established that the Neptunian arcs are concentrations of particles embedded within the narrow Adams ring. Since Voyager, the ring arcs were imaged in 1998 with NICMOS on the Hubble Space Telescope and from the ground by the Canada–France–Hawaii Telescope as described in Section 13.6. Namouni and Porco have proposed a possible solution where the arcs occupy the resonance sites of Galatea's corotational eccentricity resonance, each of 43 sites having an angular width of 8°.37. In 2002 and 2003, Imke de Pater of the University of California at Berkeley and her colleagues (de Pater *et al.*, 2005), made observations with the Keck telescope that confirm many aspects of Namouni and Porco's proposal.

Using adaptive optics techniques, de Pater *et al.* were able to see the complete Adams and Le Verrier rings. See Figure 13.7, which also shows an inset of Neptune and Neptune's moons Proteus, Larissa, Galatea and Despina. The moons have multiple exposures because of the multiple images added together. The ring arcs are shown in Figure 13.8 along with longitudinal profiles comparing the observations of Voyager, NICMOS, and Keck.

The images clearly show the moon Galatea orbiting Neptune just inside the Adams ring, and Despina just interior to the LeVerrier ring. The LeVerrier ring is about the same brightness at all longitudes, but the Adams ring is brightest near the ring arcs, as observed by Voyager in 1989. It is about three times brighter there than at the point in the ring opposite the arcs. Comparing the various observations, de Pater *et al.* show that the relative brightness at the arcs has changed dramatically. Although the overall appearance of the rings is mostly unchanged, the arcs have changed! The Keck adaptive optics observations in particular show dramatic changes in the ring arcs. Although Fraternité follows Nicholson's solution 2, all other arcs shift in location and intensity relative to it. Egalité 1 and 2 appear to have reversed in relative intensity, perhaps because material migrated between resonance sites.

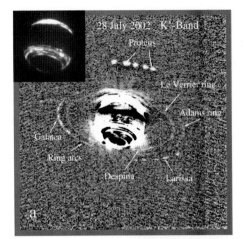
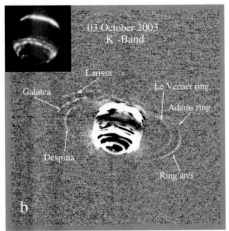
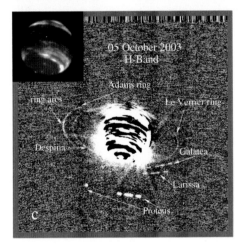

Figure 13.7 (a) Average image of the July 28, 2002 data (2.12 μm, 33 min total integration time), revealing satellites, ring arcs, and the complete Adams and Le Verrier rings. The images have been high-pass filtered by subtracting the same image median-smoothed with a width of 50 pixels. This procedure removes diffuse scattered light, and brings out small-scale features (Neptune is highly saturated in this presentation). A 1-min exposure of Neptune itself at K′-band is shown in the insert. (b) Average high-pass filtered image of the October 3, 2003 data (2.12 μm, 30 min total integration time). The insert shows a 1-min exposure of Neptune itself at K′-band. (c) Average high-pass filtered image of the October 5, 2003 data (1.63 μm, 23 min total integration time). The insert shows a 1-min exposure of Neptune itself at H-band. (From de Pater *et al.* (2005); reprinted with permission from Elsevier.)

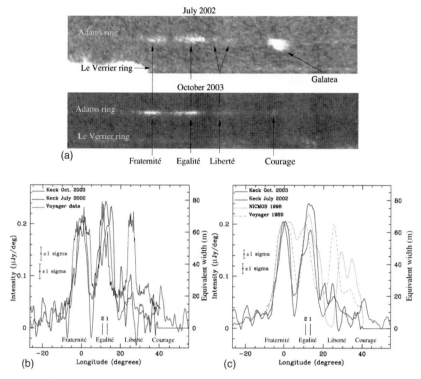

Figure 13.8 Deprojected images of the ring arcs from July 28, 2002 (upper panel; 33 min integration time) and all data in October combined (lower panel; 80 min integration time). The image dimensions are ~100° in longitude and ~1.1 × 10⁴ km in radial distance. Since Galatea (upper panel) is much brighter than the ring arcs, this moon, as well as its PSF pattern, are saturated on this display. The PSF pattern around this moon causes Galatea's odd shape. (b) Scans in longitude through the ring arcs from July 2002 and October 2003 at full angular resolution (FWHM is ~1°), together with a Voyager profile (scaled in intensity to match Fraternité) at a phase angle of 135°. The Keck data were integrated radially over the FWHM of the rings, and the intensity scale was adapted to represent the integral over the full radial extent of the rings. Equivalent width (in m) is shown on the right. Zero longitude was chosen to coincide with the center of arc Fraternité. (c) Longitudinal scans through the ring arcs as seen by Keck, Voyager (1989) and NICMOS (1998). The Keck and Voyager profiles were smoothed to a resolution of 3°. All intensities are scaled to that of Fraternité in October 2003. Equivalent width (in m) is shown on the right. Zero longitude was chosen to coincide with the center of arc Fraternité. (From de Pater *et al.* (2005); reprinted with permission from Elsevier.) For color version, see Plates section.

Courage, which is usually dim, flared in intensity to become nearly as bright as Liberté when NICMOS observed it in 1998 (Figure 13.5). In 2003, both Courage's and Liberté's brightness decreased to that of Courage in 1989. Additionally, Courage has apparently shifted 8° relative to its Voyager longitude. This would be consistent with it moving to the next corotation resonance site, which is 8.37° away in the model of Namouni and Porco.

On all the days that Keck observed, it was comforting that the position of the arc Fraternité was exactly where predicted by Nicholson's solution 2. The inferred motions therefore rule out the 86:84 corotation resonance initially predicted by Porco in 1991. The arcs Fraternité (10°) and Liberté (8°) have widths similar to that predicted by the CER. These findings are both consistent with Namouni and Porco's model, but some questions remain.

The red color of the ring arcs is consistent with their being dust produced from erosion of the moonlets. Like the F ring (French *et al.*, 2012) of Saturn, the Neptunian arcs show significant changes over decadal time periods. No quantitative theory describes such rapid evolution, although competing processes of aggregation and fragmentation (Barbara and Esposito, 2002; Esposito *et al.*, 2008), moonlet collisions (Atree *et al.*, 2012), and other stochastic processes may be operating there.

The role of random processes to renew rings and the resultant variability can explain why the multiple examples of ring systems are more dynamic than expected just a decade ago. Although the rapid processes were originally interpreted as a sign of youth (see, e.g., Esposito, 1986), they now provide evidence for continuing renewal, perhaps even for Saturn's broad rings (Esposito *et al.*, 2008). Tidally limited accretion and the efforts of moon forcing can allow clumps to grow even inside the Roche limit. Rare random events can compact the resulting clumps. This could make them gravitationally bound, and thus resistant to erosion and fragmentation. These aggregates can provide a reservoir to renew the rings when the aggregates are disrupted by an external impact. Because of the stochastic nature of these processes, it is difficult to say if some or all of the ring arcs will decay away or bounce back into existence. Naturally, some new arcs may even appear at completely new locations.

13.9 Summary

Neptune's rings have been puzzling since the first observational searches. The Voyager close-up views provided the information needed to explain the puzzles, but unfortunately not correctly. In 1998, the situation seemed hopeless: our best models were contradicted by Hubble and ground-based telescopes. With the clever proposal of Namouni and Porco, we now have hope. Their theory connects the arc mass and Galatea's orbital eccentricity but cannot be confirmed without close-up observations. We still have a disagreement between Salo and Porco as to whether the observed arcs lie *between* or *at* the resonance locations. Furthermore, the Namouni and Porco explanation predicts arc lengths of twice that predicted by Porco: 8.4°

instead of 4.2°. It is a true challenge to fit the observations into segments of this length, although the dynamics may be even more challenging than imagined now (whew!), given the multiple resonances and possible embedded moonlets.

It is clear that the understanding of the Neptune ring arcs will involve numerous stochastic elements (see Chapter 9). Furthermore, we have expended great effort to explain how such partial ring structures could persist for decades. A bigger problem is to explain how they might survive for millions of years or for the age of the solar system. Are they truly unique, random features, rarely to be repeated, or simply unusual configurations that merely represent one moment of a long random process with many possible realizations? My belief is that it is more likely the latter (see Chapter 10 on cosmic recycling). Although we can now draw back from describing Neptunian rings as "misbehaving," complete understanding eludes us still.

14

Jupiter's ring–moon system after Galileo and New Horizons

We recall from Chapter 3 that Jupiter's rings are highly transparent and contain large numbers of dust-sized particles that are most easily seen when backlit by the Sun. The first evidence that Jupiter possessed a ring system came from the Pioneer 10 and 11 particle flux measurements of 1972 and 1973, which showed regions of unexpectedly few charged particles. Much later, Voyager returned a smeared, multiply exposed image of the main ring in 1979. A relatively bright main ring circumscribes a vertically extended halo. Galileo's camera showed the gossamer ring seen by Voyager to be actually two distinct rings that are derived from Jupiter's moons Amalthea and Thebe. An artist's conception of the Jupiter rings shows clearly the intimate relation between the observed rings and their embedded satellites (see Figure 3.3). Adrastea lies just in the outer edges of the main ring; Metis is 1000 km inside, where the ring is depleted. Each of the vertically thick gossamer rings is associated with a moon on an inclined orbit: the innermost gossamer ring originates from Amalthea, and similarly the outer gossamer ring extends inwards from Thebe. The short lifetimes of the tiny ring particles require continuing replenishment, with the ring-moons as the likely source. Metis and Adrastea are about 10 km in radius, and possibly two remaining fragments from a larger precursor moon. Thebe and Amalthea are much larger, approximately 100 km in size. The gossamer rings are probably inwardly drifting particles kicked off these satellites by impacting meteoroids, as proposed by Burns et al. (1980). The outward evolution of dust from Thebe is likely to be caused by a "shadow resonance" where successive passages through Jupiter's shadow can accelerate the small particles (Hamilton and Krüger, 2008; Horanyi and Burns, 1991).

The rings' mass is insignificant compared to that of the nearby moons, allowing an almost inexhaustible source of material for the ethereal disk. The backscattered light is probably dominated by reddish parent bodies like Adrastea. Longitudinal brightness variations are seen with amplitude of 10–20%. Surprisingly, the asymmetries seen by Galileo are opposite to those seen decades earlier by

Voyager. Shawn Brooks *et al.* (2004) believe that the variations are best explained by transient clouds of debris released during mutual collisions between parent bodies, similar to Barbara and Esposito's (2002) model for brightness variations in Saturn's rings. New Horizons flew by Jupiter in 2007 on its way to Pluto. No satellites larger than 0.5 km were found, but seven elongated clumps of particles were detected. They may resemble the many elongated features seen in Saturn's F ring that I have termed "kittens" (see Chapter 11) or jets that may arise from aggregations initiated by forcing at satellite mean motion resonances.

Jupiter's gossamer rings are about 30 times dimmer than the main ring (itself quite faint). Galileo images showed that this feature detected by Voyager is actually two thin rings. A very close match between the vertical excursions of the moons Amalthea and Thebe (due to their inclined orbits) and the rings' thickness shows the moons are their sources (see Figure 3.4). Burns *et al.* (1999) explain how dust grains ejected from the surface of each moon evolve inward toward Jupiter, creating the two overlapping, wedge-shaped rings that are seen. Some small fraction might also evolve outward from Thebe, as observed. Despite some predictions, we find no direct evidence of Lorentz resonances (resonant interactions between Jupiter's magnetic field and the particles' orbits, similar to the Lindblad resonances (Chapter 6) seen in Saturn's rings; the physics of Lorentz resonances is discussed in more detail in Chapter 16). This may mean that gravitational forces may be sufficient to explain the motions of the small dust particles. Alternatively, orbital variations may smear out observable magnetic signatures.

The ring-moons appear clearly as sources of the gossamer rings. Such small bodies are likely to be the source of Jupiter's main ring, too. Collision speeds of interplanetary projectiles near Jupiter are greater than $10 \, \text{km} \, \text{s}^{-1}$ due to Jupiter's gravity. The weak gravity of small moons means their escape velocity is only tens of meters per second. In fact, escape is even easier from some parts of the inner moons, owing to their elongated shapes and their rotation. Because of the lower escape speed, Metis and Adrastea may supply the main ring more copiously than the larger Thebe and Amalthea supply the gossamer rings. The main ring probably contains many other parent bodies. Strong evidence for these is the visibility of the main ring from Earth, which indicates that the ring particles scatter significant light backward toward the Sun. Small micrometer-sized dust particles do not: they scatter light mostly forward (see Chapter 15).

In Jupiter's ring system, like those of Saturn, Uranus, and Neptune, small moons are intimately intermixed among the rings, acting as both sources and sinks for ring material. These small satellites have undergone complex histories of cratering, fragmentation, and regolith evolution. Estimates of Amalthea's bulk density yield about $1 \, \text{g} \, \text{cm}^{-3}$, which is less than for solid rock. This implies that it also is likely to be formed of unconsolidated material like a pile of rubble. It is likely that similar, smaller rubble piles continue to resupply the Jovian ring system. Impacts can then produce the debris that we see as rings.

14.1 Cassini observations of Jupiter's rings

En route to Saturn, Cassini's cameras took more than 1000 images of the Jupiter ring system, covering a range of viewing geometries and colors. These pictures are unfortunately compromised by scattered light from Jupiter so that only the main ring is visible (see Figure 14.1). Typical ring signals are only a small percentage of the background light. Images of the ring show small elongated clumps about 20% brighter than the surrounding ring. Using the Cassini data along with previous observations, Throop *et al.* (2004) find that the biggest contribution to the light scattered comes from particles about 15 μm in radius; this may be the result of size-dependent loss processes on an initially broad distribution (see below). A major conclusion from analysis of the Cassini data is that non-spherical particles fit the data significantly better than spherical particles. This is entirely consistent with the collisional source and rough conditions that these ring particles must endure. The parent bodies that supply the ring's dust are intrinsically quite red, similar to Amalthea.

14.2 Vertical corrugations

Galileo and New Horizons images show two sets of spiraling vertical corrugations in the main ring several kilometers high (see the papers by Hedman *et al.* (2011) and Showalter *et al.* (2011)). These are probably the result of a large external body hitting and tilting the rings. Showalter traces one pattern to the impact of SL9 in 1994, which left huge scars in Jupiter's atmosphere that persisted for months. Amazingly, the results of ring impacts have persisted for decades. Similar spiral features in Saturn's rings were visible above the ring surface when the sun set on the rings during the equinox of 2009, observed closely by Cassini.

14.3 Size distribution in Jupiter's main ring

Measurements of the ring's brightness at different angles and at different wavelengths strongly constrain the sizes of the ring particles. The angular dependence of the light scattered by an individual ring particle is termed its *phase function*. This function depends on the particle's size, shape, index of refraction, surface roughness, and porosity, and can only be calculated exactly for certain limiting cases. The most useful of these is the case of a uniform dielectric sphere, first calculated by Gustav Mie in 1908 (see van de Hulst, 1957). Observations of backlit rings emphasize small scattering angles since the Sun is behind and in nearly the

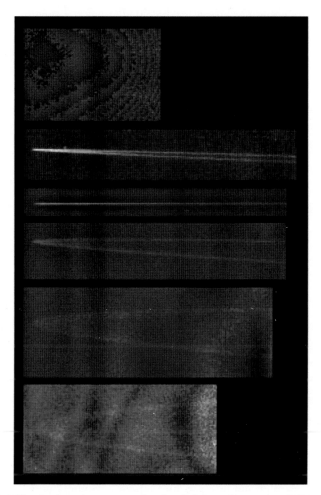

Figure 14.1 The Jovian ring as seen by Cassini from December 2000 to January 2001. The phase angles are approximately 1°, 60°, 64°, 75°, 94°, and 120°, while the ranges are approximately $270\,R_J$, $137\,R_J$, $137\,R_J$, $141\,R_J$, $162\,R_J$, and $247\,R_J$. All images are taken through clear filters, with exposure lengths of several seconds. Stray light increases as Cassini leaves the system and the rings once again appear closer to Jupiter. The spatial resolution across these images varies from 120 km at the ends to 59 km near closest approach. Images have been rescaled in brightness but not spatially. (Figure from Porco *et al.* (2003); supplementary online materials.)

same direction as the illuminated ring particles. This, in turn, selects those particles whose size is comparable to the wavelength of observation, because their diffraction lobe lies nearly in the forward direction. The typical width of the diffraction lobe (in radians) is λ/a where λ is the wavelength of observation and a the particle radius. For Galileo observations using the camera and the near-infrared mapping spectrometer, this means that particles of size 1–100 μm dominate the light seen in the pictures. Pollack and Cuzzi (1980) compared the results of light-scattering experiments for a variety of shapes to the calculations for spheres. They found the scattering behavior of real particles is well matched by calculations for spheres, for small scattering angles like those of Galileo observing Jupiter's rings. The match is improved for randomly oriented particles whose size is comparable to the wavelength of observation.

From Chapter 15, we find that the optical depth of a ring (a measure of its total capability to interact with radiation) is given by the integral over its size distribution of its efficiency to scatter radiation:

$$\tau(\lambda) = \int_{a_{\min}}^{a_{\max}} n(a)\pi a^2 Q(a, \lambda)\,da \qquad (14.1)$$

Here $n(a)$ is the vertically integrated number of ring particles per unit area in the size range from radius a to radius $a + da$. The efficiency for extinction $Q(a, \lambda)$ can be calculated from Mie's scattering theory. The most common assumption is that the size distribution is a function of some power of the radius, that is,

$$n(a) = C\left(\frac{a}{a_0}\right)^{-q} \quad \text{for } a \text{ between } a_{\min} \text{ and } a_{\max} \qquad (14.2)$$

where a_0 is a reference size, C is an appropriately chosen constant, and q is termed the (power-law or spectral) index of the size distribution. For realistic distributions, we find q positive; in such a distribution, smaller particles are thus more abundant. This power law is probably the simplest realistic distribution; laboratory simulations of collisional debris give power-law distributions; and a wide variety can be studied by varying the single parameter q. Because the size distribution in planetary rings is probably the result of multiple collisions, determining this spectral index from observations constrains their collisional history. The combination of the variation of ring brightness with angle (seen by Galileo's camera) with its variation with wavelength (observed by the spectrometer) gives a powerful method to determine the size distribution. This was the topic of Shawn Brooks' (2003) thesis at the University of Colorado (see Brooks *et al.*, 2004).

Brooks found that a more complicated size distribution than a simple power law is needed to explain the Galileo observations. He employed a combination of two power-law distributions, with different index above and below some breakpoint in radius. This gives

$$n(a) = C_1\left(\frac{a}{a_0}\right)^{-q} \quad \text{for } a \leq a_{\text{brk}} \qquad (14.3)$$

$$= C_2\left(\frac{a}{a_0}\right)^{-(q+dq)} \quad \text{for } a > a_{\text{brk}} \qquad (14.4)$$

Since the distribution is continuous at $a = a_{brk}$

$$C_2 = C_1 \left(\frac{a_{brk}}{a_0} \right)^{dq} \qquad 14.5$$

Even this distribution is surely an oversimplification of the actual size distribution. Nonetheless, it captures information about the real size distribution that is seen by the multiple Galileo observations. The "broken" power-law size distribution reproduces the Galileo Near Infrared Mapping Spectrometer (NIMS) spectral observations much better than the best fits using a single power-law index. In Brooks' model, the Galileo camera observations determine the value of the power-law index q, while dq is constrained more by the NIMS observations in the infrared. His results suggest that the breakpoint in radius is at about 15 μm. Below this, the size distribution resembles a power law with index $q \cong 2$. This is consistent with Showalter *et al.*'s (1987) analysis of the Voyager camera observations. For particles larger than 15 μm, the distribution is steeper, with $dq \cong 3$. This same distribution has also been used by Throop *et al.* (2004) to explain the Cassini observations of Jupiter during its distant flyby of 2000. Thus, the Voyager, Galileo, and Cassini data yield a consistent picture of Jupiter's main ring.

What are the processes that could produce such a ring? The Galileo observations provide strong support for the idea that the ring material is detritus continually supplied from small moons among the ring. One possibility is that the size distribution we see is exactly the one produced from the moons' collisions with each other and with meteoroids. We could then relate it directly to the bombardment process. A second possibility is that the distribution results from the combination of both production processes and of loss processes of ring particles. In the steady state, these would exactly balance. However, loss processes depend on size (see Chapter 16): this means that the observed size distribution will always be different from the production size distribution. Most loss processes preferentially remove small particles. For a power-law distribution, this means that the index q decreases with time. Alternatively, Canup *et al.* (1993) also discuss processes that could steepen the initial size distribution and would thus increase the value of q.

This discussion assumes a steady-state balance is achieved. Because ring particles are created by stochastic events, this might be overly optimistic. Continuing production events will make the observed distribution episodically resemble more the production distribution; as time passes, it will steadily evolve to approach the distribution where balance determines the characteristics of the observed size distribution. Furthermore, the balancing processes proceed at different rates for different-sized particles. This means that some parts of the size distribution (particularly the smaller ones) will be in balance, while it takes much longer for the larger particles to reach equilibrium.

This may provide a natural explanation for the "broken" power-law distribution. If the majority of particles in the Jovian main ring were created by the catastrophic break-up of a large progenitor body in the recent past, and there has been little replenishment of ring material, the size distribution has been continually evolving since. The smaller particles are quickly lost and a_{brk} (the size of the particle at which the power-law index changes) increases as time passes. For particles below this size, equilibrium has been achieved and the power-law index has value q, representing balance between gain and loss. For larger particles, the size distribution is steeper, with its larger values of q representing the initial distribution produced at the catastrophic disruption that created the ring.

In this latter case, the location of the break in the derived size distribution would give an approximate age for the ring. For the value of $a_{brk} = 15\,\mu m$ from Brooks *et al.* (2004) and the loss rates from Mihaly Horanyi and Tom Cravens (1996) this gives an age of only 1000 years. Even if the slower evolution rates from Burns *et al.* (1999) are used, the ring would still be a relatively young 10^6 years old. Even if the rings are more ancient, the small particles in them have short lifetimes of only 100–1000 years.

Can we really conclude that the main rings of Jupiter are only 1000 years old? Unfortunately, this would be premature. Because of uncertainties in the size distribution, and dispute over the effectiveness of various loss mechanisms and the rates at which they proceed, we are unable at this time to decide between the various explanations for the history and evolution of the Jovian main ring. However, some of the same processes are likely to be active in Saturn's rings. Thus, better understanding of Saturn's thinner rings E, F, and G sheds some light on this controversy. Cassini observations of the faint Saturn rings clearly show rings created by erosion of small moons (Hedman *et al.*, 2009). Nonetheless, models with either episodic, catastrophic, rapidly evolving rings, or slower, more sedate models with rings sculpted by subtle magnetic resonances are both consistent with our observations of Jupiter's ethereal rings.

14.4 Summary

It is not surprising (Showalter, 1989) that Galileo, Cassini and New Horizons observations have expanded our understanding of Jupiter's rings. The Galileo data clearly show that the gossamer rings are derived from the small moons Thebe and Amalthea. This confirms the general expectation from the Voyager observations that the tiny particles in the rings are created from material knocked off small moons. The main ring probably contains similar parent bodies. The Galileo observations were from a variety of viewing geometries and over a range of wavelengths. A model of the size distribution as a single power law cannot fit all observations simultaneously. Brooks *et al.* (2004) were able to match the Voyager and Galileo observations using a more complicated "broken" power-law particle size distribution. This model can also match the Cassini observations. Such a size distribution could result from evolution of the ring particles created from a recent disruption of a small moon. If so, the current Jovian main ring might be only 1000 years old. Unfortunately, data and models are too sketchy to make this conclusion solid. New Horizons shows clumps in the rings, possibly associated with embedded objects, which may serve as "seeds" for the aggregations. Impacts into the Jovian ring occur every decade or so, leaving characteristic spiral corrugations. Together these observations show a combination of internal and external processes creating and sculpting the ring system.

15

Ring photometry and spectroscopy

With the exception of Cassini's Cosmic Dust Analyzer (CDA) that detected individual particles in Saturn's E and G rings, we have never actually touched a ring particle, and our observations of rings are made from some distance. We have achieved the majority of our present understanding of their composition, behavior, and history by remote sensing. This justifies a short discussion of how visible light or other electromagnetic radiation interacts with rings, and what we can learn from measuring the brightness of light scattered by the rings: literally ring photometry. The basic approach is to consider the rings to resemble an atmosphere of scatterers; this is the same technique used to estimate the brightness of a cloud or haze. For the atmosphere case, standard solutions exist, for example by James Hansen and Larry Travis (1974), William Irvine (1975), and Gary Thomas and Knut Stamnes (1999). Cuzzi *et al.* (1984) give a detailed description of the application of these radiative transfer methods to planetary rings. This standard approach assumes the ring particles are distributed in height and well separated both in terms of the particle size, and relative to the wavelength of light used for their observation. For some cases, these assumptions may not be met: for example, for some observations at the longer radio wavelengths and for the densest parts of the rings of Saturn. Furthermore, much of the scattering can be among the rough elements of a single ring particle, as well as between particles. For a detailed discussion of the coherent behavior of light scattered backwards by a rough surface, see the work of Michael Mishchenko (1993), for example.

15.1 Photometry

If we assume homogeneous spherical ring particles, their interaction with radiation depends only on their index of refraction and size compared with the wavelength; this can be calculated for pure dielectric materials by the standard methods first developed by Gustav Mie, called "Mie scattering." Such calculations show that the effective area for interaction is generally different from the actual physical cross-section $A = \pi r^2$. The difference is specified by an efficiency factor Q, so that the effective cross-section is $Q\pi r^2$. For more details, see the excellent book by van de Hulst (1957), *Light Scattering by Small Particles*. Although ring particles are generally *not* spherical, the efficiency Q is not strongly dependent on particle shape but more on its size relative to the wavelength of observation, and on the particles' optical properties. The angular distribution of light scattered by a particle is called its *phase function*. Standard techniques are available to calculate this, and also the multiple interactions of particles with radiation, so that the total reflected light may be predicted for a flat layer of ring particles possessing a distribution of size ranging from some smallest to largest particle.

For particles much smaller than the wavelength of observation, scattering is very inefficient, increasing rapidly for particles about the size of the scattered light, and for large particles the efficiency for scattering approaches the physical optics limit $Q = 2$. This value of 2 may seem somewhat surprising: the fact that an obstructing particle removes radiation equivalent to that falling on twice its area is known as Babinet's paradox (or principle). In this limit, a particle diffracts an equal amount of radiation to that which directly strikes its physical cross-section. Thus, it removes twice as much energy from an incident beam of light.

For larger particles, the diffracted light is concentrated in an increasingly narrow cone about the initial, forward direction of impingent light. The typical angular width of the diffraction cone is λ/D, where λ is the wavelength of the light, and D the particle diameter. Thus, the phase function becomes more strongly forward peaked as particles become larger or if they are observed at shorter wavelengths. For particles much larger than the wavelength of observation, the forward lobe becomes so narrow that it is indistinguishable from the incident beam of light striking the ring particle. This has caused some confusion; see Cuzzi (1985) for a more detailed discussion.

However, in the special case of a radio occultation, the slight Doppler shift of the diffracted light is detectable; in this case the factor of 2 does not appear, and this means that great care is needed in comparing the results from different occultations. Sometimes a mysterious factor of 2 appears: it is just the value from Babinet's paradox.

In interpreting the light from planetary rings, we must keep in mind that we see two important effects. The first is the scattering by an individual ring particle, or from a small volume element containing an ensemble of particles in a range of sizes. The second is the collective effects of multiple scattering among the particles and of transmission into and out of the system of scatterers. Because of the optical effects of interaction between light and matter, the smallest particles tend to scatter light in all directions, the larger ones scatter light mostly backwards. For example, the surface of the Moon is highly backscattering, with a peak reflection when the light is scattered directly backwards at the time of full moon. The largest particles in Saturn's main rings behave similarly. The smallest ones are much more isotropic scatterers. Further, at some small angles of scattering, their diffraction lobes may be observable, causing their reflection to brighten significantly as they are backlit.

The collective effects can be treated by using standard routines developed for planetary atmospheres. Care must be taken if the particles in a planetary ring are not actually distributed in space like dust in a planetary atmosphere, for example if the rings are very flat, as Saturn's B ring is. This situation, called a *monolayer*, yields a somewhat different ring brightness where the collective effects are much reduced. We have good evidence that this situation may hold in Saturn's B ring and in the Uranian rings (Zebker *et al.*, 1983; Gresh, 1990).

In addition to scattering sunlight, ring particles (although cold) emit thermal radiation. This thermal emission is an important diagnostic and the Cassini spectrometers have observed the ring particles as they enter and leave Saturn's shadow. Individual particles warm in sunlight and cool off in shadow. Further, warm particles radiate on others. Measuring the ring temperature of particles entering and leaving Saturn's shadow constrains their physical properties, such as size, shape, rotation, porosity, and thermal inertia; such observations can also determine where the ring system is many particles thick or more like a monolayer.

Salo and Karjalainen (2003) have studied how the dynamics of the ring particles determine the photometric properties we observe remotely. They note that dynamical models for the collisional state in planetary rings suggest rings to be fairly flattened, with vertical thickness well below the observational upper limit set by Voyager of about 200 m (Lane *et al.*, 1982, and similar values from Cassini (Jerousek *et al.*, 2011)). The rings are more likely only tens of meters thick, as indicated by analysis of density wave damping. This means that the densest parts of rings have particles close enough together that 10–30% of the total volume is occupied by particles. Thus, the real rings lie somewhere in between the two states of many-particles-thick and monolayer discussed above. For such highly flattened systems, the mutual gravitational attraction between particles is significant. This causes the particles to clump together, forming inhomogeneities similar to the wakes raised by a boat speeding through water. In these wake regions, the local compaction is even higher.

Salo and Karjalainen forsake the framework of classical radiative transfer discussed earlier that assumes that the rings are a thick, homogeneous layer many particles thick, with a small volume filling factor. Such models, for example by Esposito and Lumme (1977), successfully explain the variation of ring brightness over a range of viewing geometries that is seen as the apparent tilt of Saturn's rings changes with Saturn's seasons. This is known as the *tilt effect*. Instead, Salo

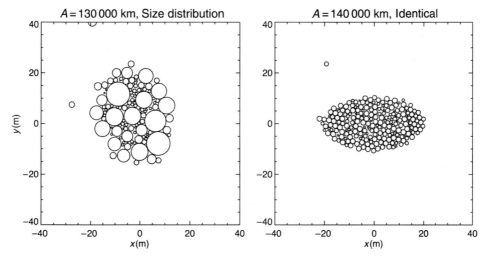

Figure 15.1 Particle accretion in Saturn's rings. Comparison between simulations with a power-law size distribution ($dN/dr \propto r^{-3}$ for $0.50\,\text{m} < r < 5\,\text{m}$) and with identical particles. A slice of the simulation system through the equatorial plane is shown. Note that the experiment on the left corresponds to $a = 130\,000\,\text{km}$, while for that on the right, $a = 140\,000\,\text{km}$. In both cases, the surface density $\sigma = 480\,\text{kg}\,\text{m}^{-2}$. (From Salo (1995); reprinted with permission from Elsevier.)

and Karjalainen developed a new model based on tracing the rays of sunlight in a planetary ring by drawing random variables at each scattering to determine the further propagation of each photon: this is a Monte Carlo method. Using 10^4–10^5 simulation particles in their dynamical analysis, they can predict the photometric consequences of different physical regimes. This allows us to constrain the ring dynamics by remote observations. In Figure 15.1, we show a number of possible states for which they calculate the ring brightness behavior. Applying their method to Saturn's rings, they find that the moderately flattened, large-density models predicted by dynamical simulations are consistent with both the seasonal tilt and opposition photometric behavior of Saturn's brightest ring, ring B.

An alternative explanation for the opposition effect is that it is due entirely to the individual particles: their rough surfaces create a "coherent" backscatter that reflects sunlight sharply backward. The opposition brightness increase would then be due to constructive interference of light from small regolith grains on the ring particles' surfaces (Mishchenko, 1993). In this case, the collective effects of many particles are insignificant. The present data set for Saturn's rings is not good enough to allow us to decide between these alternatives. Cassini observations of the rings' opposition effect provide some support for both the models (see Hapke *et al.*, 2006; Degiorgio *et al.*, 2011). Estelle Deau (2012) argues that both mechanisms may contribute to the observed brightening at opposition.

The ability to match the observations arises from the spread in sizes of the ring particles: the smaller particles have a thicker distribution because of their higher velocities (see Chapter 5). These smaller particles form a layer several times larger than that formed by bigger particles; their volume density is thus much smaller. The combination explains the brightness variation as Saturn's rings are observed near Saturn's opposition (when the Sun shines from directly behind the observer, like viewing full moon from the Earth). For the same reason, at the seasons when the Sun and Earth are at high elevation (Saturn summer and winter), the larger particles near the ring plane are lit by the Sun and also visible to ground-based observers. At the other seasons, the lighting is much more oblique and Earth observers are seeing mostly the more extended smaller particles. This explains the tilt effect, which was earlier explained by increasing multiple scattering in a many-particle-thick layer (see, e.g., Esposito and Lumme, 1977). Thus, models with a moderate central plane filling factor of about 20% can explain both the opposition and tilt effects provided the particle-size distribution is taken into account. It is amazing that our remote observations are able to test both the sizes and dynamics of the ring particles, too small to be seen individually.

15.2 Spectroscopy

The spectra of ring particles give information on their composition and evolution. Ice and mineral band absorptions are common in the near-infrared (see Clark *et al.*, 2013b). Water ice shows a strong absorption edge at about $1650 \, \mathring{A}$ (Bradley *et al.*, 2010).

This spectroscopic information is complicated by the fact that the ring particles span a range of sizes, and they tend to aggregate, with the larger particle's surface probably covered by a regolith of smaller particles. The composition changes owing to radiation striking the particles, generally darkening their surfaces. The surface textures are changed by collisions, aggregation, and erosion. Micrometeoroids erode and also pollute the ring particles. Nonetheless, analysis of their reflection and emission spectra strongly constrain their composition.

In general, particles in planetary rings are similar to the nearby moons. Saturn's rings are predominantly water ice, Uranus' rings are dark, and Jupiter's are derived from nearby Thebe and Amalthea. Although Saturn's rings are dominantly crystalline water ice, they are visually red, showing some contamination by non-icy material (see, e.g., Cuzzi *et al.*, 2009).

Color variations across Saturn's rings (Cuzzi *et al.*, 2009) may indicate varying composition, including effects of the interplanetary dust that bombards them and darkens the particles. Saturn's ring particles have rough, irregular surfaces

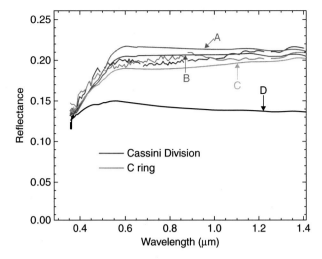

Figure 15.2 Spectra of laboratory analogs compared with Cassini VIMS spectra of Saturn's rings. Laboratory spectra at temperatures 84–90 K of ice mixtures with nano-hematite and other possible absorbers (A, B, C, and D) are shown. (From Clark *et al.*, 2013b.) For color version, see Plates section.

resembling frost more than solid ice. There is a good indication that the particles are less dense than solid ice, supporting the idea of ring particles as temporary rubble piles. These slowly spinning particles collide gently with collision velocities of usually just millimeters per second.

A recent suggestion is that the red color of the rings results from iron oxidized to hematite (Clark *et al.*, 2013b). Nano-hematite (very fine grained hematite Fe_2O_3) is a strong UV absorber that matches the ring spectrum (Figure 15.2). The oxidation may be the result of chemical reactions with the ring atmosphere (which contains the water products H_2O, OH, O, H, H_2, and O_2 and their ions; see Johnson *et al.* (2006)). The atmosphere over the rings is dominated by molecular oxygen (O_2). There is no spectral evidence for silicates in the ring particles (Clark *et al.*, 2013b).

15.3 Summary

Just as the physical properties of the ring particles determine their dynamics, the optical properties determine how they reflect and emit light. Numerical simulations of the light scattering are based on the composition, reflectivity, abundance, and volume density of the ring particles. Individual ring particles each have their own scattering behavior, characterized by the phase function. Collective effects of multiple scattering and mutual shadowing determine the ring brightness variation at different observing geometries. One approach to solving this problem is to follow the individual histories of a large number of photons. The stochastic process of the collective effects is simulated by following in detail the path of each individual photon through successive optical interactions with the ring particles. The new direction is determined by drawing a random variable: this is one example of the application of the Monte Carlo method to radiative transfer.

16

Dusty rings

Particles of a size of about 1 μm in planetary rings are often described as dust. Rings with significant populations of such small particles are accordingly described as dusty. This dust is of interest for several reasons. First, such small particles are subject to additional forces beyond gravity, and thus their behavior is more complicated and often significantly different from that of the larger particles in planetary rings. Second, these additional processes can lead to very rapid evolution, and thus short lifetimes. This requires a corresponding production of dust to balance these losses. Further, because small particles have a size similar to the wavelength of visible light, they interact strongly with it and distinctly differently from macroscopic particles (see Chapter 15). The phenomenon of diffraction causes dust particles to appear especially bright when backlit. Spacecraft to the outer planets attain a perspective of looking back toward the Sun that is not available to us Earthbound astronomers. The dusty rings light up in these views and provide unexpected, spectacular views of rings illuminated from behind. Naturally, the dusty rings and also the dust within a ring are most visible in these views. Figure 16.1a, b compares the Uranian rings seen in backscattered and forward-scattered light. The forward-scattered light shows a new population of dusty rings among which the classical rings observed from Earth are entirely undistinguished members.

16.1 Forces on dust

All ring particles, including dust, orbit the central planet: they clearly feel its gravitational attraction. The gravity of small moons also affects the ring particles, particularly at gravitational resonances or at very close proximity (see Chapter 6). The gravity of the particles themselves is evident in their collective behavior, as in the observed waves in Saturn's rings. Gravity affects small and large particles alike: dust is subject to these same forces of gravity. However, a small particle can acquire a non-negligible charge, so that electromagnetic forces become significant.

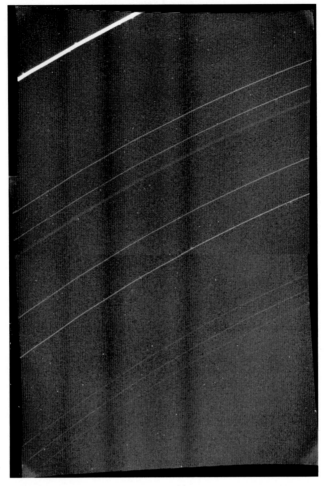

Figure 16.1a Mosaic of two low-phase-angle (21°) high-resolution (10 km per pixel) Voyager 2 images of the rings of Uranus. The planet's nine narrow optically thick rings are clearly visible, and the very narrow moderate optical depth λ ring is marginally detectable. (NASA/Voyager 2, PIA00035; courtesy NASA/JPL.).

Further, some forces act on the surface of a particle, like solar radiation pressure. The smaller the particle, the larger its area is compared with its mass: accordingly, these forces become more important for dust.

A dust particle is thus subject to the planetary oblateness, solar radiation, solar gravity, and electric and magnetic forces. In the charged plasma surrounding a planet, the dynamics of small charged dust particles is strongly influenced by electromagnetic forces, drag, and radiation pressure; the size of the effect depends

Figure 16.1b High-phase-angle (172°) Voyager 2 view of the Uranian ring system. The forward-scattering geometry dramatically enhances the visibility of the micrometer-sized dust particles. The streaks are trailed star images in this 96-s exposure (NASA/Voyager 2, PIA00142; courtesy NASA/JPL-Caltech.)

on the particle size, so that different-sized dust particles may exhibit different behavior. For dust smaller than 0.02 μm, the electromagnetic forces exceed those of planetary gravity: dust particles of this size are actually ejected from Jupiter and observed in interplanetary space (Grün *et al.*, 1994). This process has been predicted (Horanyi, 2000) and now observed at Saturn as well (Kempf *et al.*, 2005). The relative magnitude of the various forces on dust particles in the Jupiter system is shown in Figure 16.2. The forces also depend on the distance from the planet and the planet's distance from the Sun as well as the charging history of each dust grain (Horanyi, 1996). These combined forces can lead to very complicated trajectories. For the Jupiter system, this collective behavior resembles the twirling skirt of a ballerina (see Figure 16.3). Systematic variations in charge, which are experienced by grains on highly eccentric orbits, can significantly affect the particle trajectories.

Sometimes these forces can interact with interesting effect. In the region of Saturn's E ring, particles with size about 1 μm experience a periapse precession rate due to electromagnetic forces that just balance the effect of Saturn's oblateness.

Planetary Rings

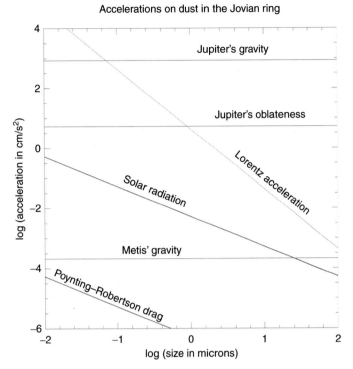

Accelerations on dust in the Jovian ring

Figure 16.2 Forces on a dust grain (electric potential $=+5$ volts, Poynting–Robertson efficiency $Q_{pr}=1$, $\rho=2.4\,\mathrm{g\,cm^{-3}}$) in the main Jovian ring as a function of particle radius. These forces depend on distance a from Jupiter in the following ways: Jupiter's gravity (a^{-2}), Jupiter's oblateness (a^{-4}), Lorentz acceleration (zero at synchronous orbit at $a=2.25\,R_J$, then increasing absolutely in either direction away from it; here we plot it for circular orbits at $1.8\,R_J$), solar radiation pressure (a^0), and Poynting–Robertson drag ($a^{-0.5}$ near Jupiter). Here Metis' gravity (unimportant away from the ring-moon) is estimated for $a \sim 1500$-km approach to the 20-km satellite. (From Burns *et al.* (2004); reprinted with the permission of Cambridge University Press.)

This approximate cancellation allows the dust particle orbits to develop large eccentricities. For this narrow range of particle size, the large eccentricities mean that the dust can impact Saturn's nearby moons like Enceladus with a high relative velocity – high enough to blast yet *more* dust particles into space. The yield could possibly be large enough that the E ring is significantly self-sustaining. We now know that the E-ring particles are produced by the eruptions of Enceladus' geysers (see below). Cassini observations show little contribution in the E ring from dust produced by bombardment of Enceladus.

Electromagnetic forces are particularly significant for Jupiter's rings, which contain many small particles and are located close to the planet, where Jupiter's

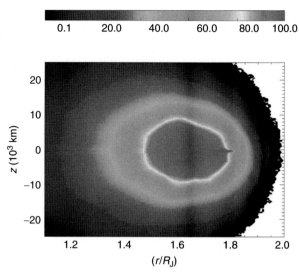

0.1   20.0   40.0   60.0   80.0 100.0

Figure 16.3 A model of the brightness produced by a dynamically derived distribution of 0.5-μm-radius grains in the vicinity of Jupiter's ring. The size is chosen to best reproduce the halo's structure. All of these grains were started uncharged with a Kepler orbital rate, assuming a uniform surface density of parents through the main ring. The grains evolve rapidly (typical lifetimes of only 20 days), under stochastic charge variations, driven by ionospheric plasma, as in the model of Horanyi and Cravens (1996). Note the brightness transition at $1.8\,R_J$, and the small wing outward, produced by short-lived grains that have larger eccentricities. (From Horanyi *et al.* (2004); reprinted with permission from Elsevier.)

own magnetic field is quite large. At certain locations, the frequency at which the particle feels the magnetic field is commensurate with its orbital frequencies. These Lorentz resonances are analogous to the gravitational resonances seen in other rings. They may cause observable effects, like the puffing up of the halo around Jupiter's main ring. The orbits of small particles become more inclined at the resonance and thus the halo is formed. In Jupiter's rings (unlike Saturn's), both collisions and self-gravity are unimportant. For Jupiter, the effects on dust can be much more evident (see Horanyi, 1998).

16.2 Dust lifetimes

Small particles are more quickly lost from planetary rings. Any gas from the planet's atmosphere causes a friction that is felt more strongly by the smaller particles. A similar drag is due to collisions with charged particles in the planet's magnetosphere. This is called plasma drag. Inside of the distance of the synchronous orbit, where the orbital period is the same as the planet's rotation, ring particles experience a headwind: their orbit speeds exceed that of the planet. The plasma trapped in the magnetosphere is rotating at nearly the planetary rotation rate, because the individual ions and electrons are gyrating about the magnetic field lines, whose rotation rate is that of the planet. Similarly, particles rotating outside synchronous orbit experience a tailwind. Their changing orbits due to drag thus lead to a depopulation of the ring region.

Another drag force is the *Poynting–Robertson effect*. This is a relativistic effect due to the dust particles' motion relative to the Sun. The reradiation of absorbed solar energy is emitted in a frame moving with the particle and not stationary with respect to the Sun. Orbits collapse under Poynting–Robertson drag in approximately the time it takes for a particle to absorb the equivalent of its own mass in radiation. For a 1-μm dust grain, this is about 10^5 years.

Mass deposited from micrometeoroids onto the dust grains also changes their orbits and slows them. In every case, dust particles will leave the ring region and may end up reaching the top of the planet's atmosphere. At that time, the dust is decelerated and enters the planet's atmosphere with the flash of a meteor.

In addition to being swept out of ring systems by the orbital evolution described above, dust particles are simultaneously destroyed by their inhospitable environments. Pure water ice grains of 1 μm orbiting Jupiter or Saturn would evaporate into the vacuum of space in 10^3 to 10^6 years. The impacts of the charged particles trapped in the magnetosphere and micrometeoroids from space gradually grind down the dust grains in an even shorter time. This process is called sputtering. Similarly, small moons sweep up and collect whatever dust strikes their surfaces. Such short lifetimes require that dust in rings must be continually recreated.

The likely source of this dust is small moons intermingled with the ring system. These are thus the *parent* bodies for the ring. They are continually bombarded by interplanetary micrometeoroids. Most circumplanetary dust is generated by micrometeoroid impacts onto the parent bodies. An individual impact can produce 10^4 times the impacting mass in a cloud of dust. Small parent bodies 5 km to 10 km in radius are very effective sources, as shown by Thebe and Adrastea in the Jovian system. Most of these parent bodies are too small to have been seen individually by spacecraft cameras: we detect them by the debris ring of dust they leave around a planet. Dust can also be created if aggregate bodies collide at a high enough velocity to release dust (Colwell and Esposito, 1990a, b; Barbara and Esposito, 2002).

16.3 Dust clouds

The Galileo spacecraft took photographs of the Jovian gossamer rings showing that they are clearly derived from small parent moons (see Figure 3.4). Galileo also detected directly the dust particles near Jupiter's largest satellites. During its orbital tour about Jupiter, the Galileo spacecraft had a total of 31 targeted flybys of all four Galilean satellites. During many of the encounters, the number of dust grains hitting the spacecraft peaked near closest approach (Krüger *et al.*, 2004), showing dust near Europa, Ganymede, and Callisto. This indicates long-lived, steady dust clouds surrounding the moons: the moons themselves are the sources of continuous ejection of debris via bombardment of their surfaces by interplanetary meteoroids. The majority of the detected grains have a size around 1 μm. These dust clouds are too thin to be detected remotely. Remarkably, a total of only 65 cloud particles were detected during the eight Galileo flybys of Europa.

Models of these dust clouds consider both the initial production of dust knocked off the moon's surface and the resulting steady-state distribution near the moon (Krüger *et al.*, 2004). The detailed models, although simplified, provide a satisfactory fit to the observations. Interestingly, about one half of the particles observed by Galileo's dust detector are not escaping but are on ballistic trajectories: within minutes to hours, they will fall back to the moon's surface. The total mass in such a cloud is 1–100 tons. Only a small fraction ($\leq 10^{-3}$) actually escapes the satellite (Krüger *et al.*, 2004).

16.4 More dusty Jupiter rings

What happens to this dust? Naturally, it also forms a tenuous ring around Jupiter, thinner and more transparent than the photographed main Jupiter rings (Thiessen-husen *et al.*, 2000). These particles escaping the moons are supplemented by captured interplanetary grains. The Galilean ring of material formed by these dusty grains escaping from the Galilean moons is far too tenuous for optical detection.

Even beyond the moons, yet another dust ring surrounds Jupiter at a distance of $50 R_J$ and more! The absolute number density (10 dust particles per km^3) is 100 times less than the density in the Galilean ring near Europa's orbit. This dust is probably magnetospherically captured interplanetary dust (Colwell *et al.*, 1998). Capture is especially efficient for particles of dust just somewhat smaller than 1 μm. It also includes dust from comet Shoemaker–Levy 9, which crashed into Jupiter in 1993, and dust produced from impacts onto the small, irregular outer moons of Jupiter. Radiation and electromagnetic forces dominate the motions of these small grains. The gravity of nearby satellites thickens the ring.

The major loss mechanisms for this tenuous dust ring are reimpacting onto a parent moon, colliding with Jupiter, or escaping the Jupiter system. Lifetimes are merely a century or less. Other loss mechanisms (plasma sputtering, micrometeoroid bombardment, sublimation, plasma, and Poynting–Robertson drag; see Section 16.2) are yet slower.

The bombardment of the outer Jupiter satellites creates ejected dust that fills an immense spatial volume. This outer ring is one of the largest but most diffuse of the planetary rings.

16.5 Saturn's ethereal dusty rings

Cassini has observed Saturn's diffuse rings with remote sensing and direct dust detectors (see the review by Horanyi *et al.*, 2009). This includes faint rings in and around the rings, spokes (see Section 7.3), narrow outer faint rings, and Saturn's E ring, which is derived from small ice grains erupted from Enceladus' cryovolcanic geysers. These rings (except for the E ring and spokes) are all collisional debris. Surprisingly, the D ring has changed since Voyager observed it 30 years ago. The faint rings are also time-variable and some have elliptical orbits aligned with the Sun.

The diffuse rings are composed primarily of particles smaller than $100\,\mu$m. As described above, the small sizes make them sensitive to non-gravitational forces. Figure 16.4 shows the rings as seen in a spectacular view looking back toward the Sun while the Cassini spacecraft was in Saturn's shadow on September 15, 2006. The dusty rings are all clearly evident. The D ring is interior to the main ring. A few spokes are hovering over the rings. Beyond the F ring (the brightest in the image) is a series of narrow dusty rings (the G ring is the brightest). The broad E ring extends from Mimas orbit outwards.

Before Cassini, our knowledge was based mostly on Voyager observations supplemented by observations from Earth of the ring plane crossings of 1995–6, when Earth passed three times through the plane of Saturn's rings. Cassini has now detected dusty material throughout the ring system. Figure 16.5 shows bright bands in Saturn's D ring. Generally, the parts of the D ring closer to Saturn have more small particles. A vertical corrugation is likely to be the result of a recent impact on the ring (Hedman *et al.*, 2007a). Comparing observations at different times, the ring was probably struck by an external impactor in 1984! (See Figure 16.6.)

G ring

Several relatively narrow rings lie between the F ring and the E ring. The brightest is the G ring, between 165 000 and 175 000 km from Saturn. It has a sharp inner edge and a diffuse outer boundary. Near the inner edge at 167 500 km, a bright arc extends over roughly 60° of longitude, observed multiple times by Cassini. The measured motion of this arc determines that it is at the 6:7 corotation eccentricity resonance with Mimas (Hedman *et al.*, 2007b) and thus it resembles the Neptunian

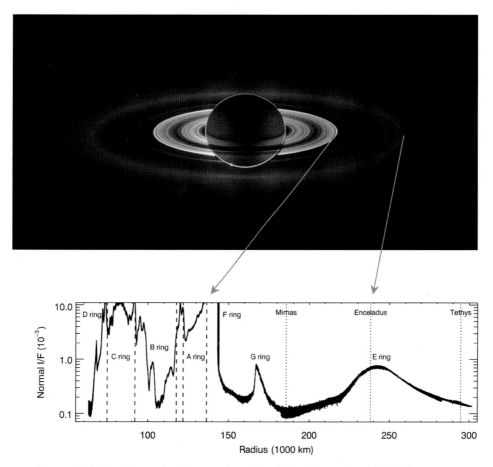

Figure 16.4 Top: A mosaic of images from Cassini on September 15, 2006, same as Figure 7.4. Because of the viewing geometry, the small particles appear especially bright. Bottom: Ring brightness observed by Cassini, plotted against distance from Saturn. (From Horanyi *et al.* (2009); reprinted with permission from Springer-Verlag.) For color version, see Plates section.

arcs discussed in Chapter 13. When the Cassini spacecraft flew through this arc, it detected many fewer charged particles, just as Pioneer 11 did in 1979. This suggests a total mass equivalent to a small moonlet 100 m across, and a small moonlet Aegeon was subsequently found embedded in this arc! (See Figure 16.7.) Probably this is just the largest of a population of particles from 1 m to 100 m across as originally suggested by Cuzzi and Burns to explain the Pioneer observations (see Chapter 11).

Gravity causes the larger particles to be held in the resonance with Mimas, but dust produced by their collisions is subject to more plasma forces (see earlier) and

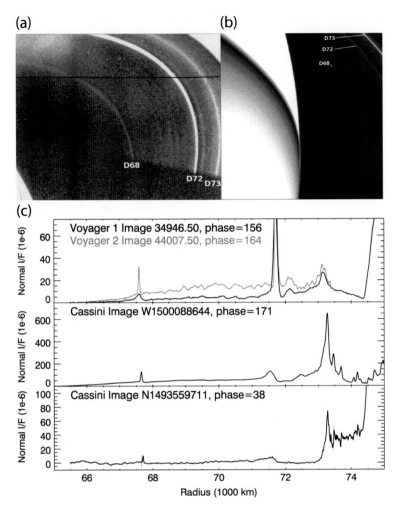

Figure 16.5 D ring. (a) Voyager image 1 (34946.50) taken at a phase angle of 156°. The three brightest bands were called D68, D72 and D73. The horizontal black line is a data dropout. (b) The same region imaged by Cassini (W1500088644), at a phase angle of 171°; an over-exposed Saturn fills the image's left half. (c) Radial brightness profiles derived from the two images above, a Voyager 2 frame (#44007.50 at 164° phase) and an additional Cassini image (N1493559711) taken at a phase angle of 38°. The vertical dotted lines mark the positions of D68, D72 and D73 from left to right. (From Horanyi *et al.* (2009), after Hedman *et al.* (2007c); reprinted with permission from Elsevier.)

leaks out to form the rest of the G ring encircling Saturn. Once again, very similar to Neptune's ring arcs! Other faint rings resembling Jupiter's have also been found associated with the small moons Methone, Anthe, and Pallene (Hedman *et al.*, 2009). (See Figure 16.8.)

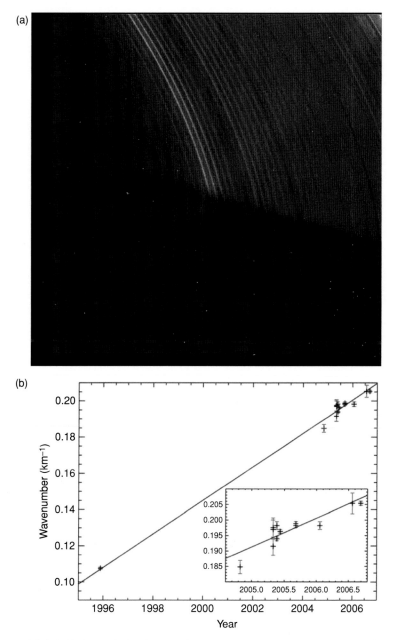

Figure 16.6 A high-resolution image shows a structure with a wavelength of about 30 km in the outer D ring. These regular brightness variations are gradually changing in separation (bottom), indicating that an object hit the ring in 1984. (From Horanyi *et al.* (2009), after Hedman *et al.* (2007c); reprinted with permission from Springer-Verlag.)

Planetary Rings

(a)

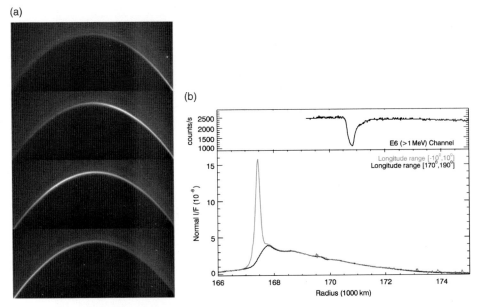

Figure 16.7 (a) Images of the G ring arc obtained on September 19, 2006, at 12:37, 13:11, 13:44, and 14:18 UTC from top to bottom. The bright arc moves from right to left through the field of view. (b) Top: The dropout in charged-particle flux detected during Cassini's passage over the arc region on September 5, 2005. The radial scale corresponds to the equatorial distance of the unperturbed magnetic field lines that thread Cassini at the time of the observation. Bottom: Average (offset-subtracted) radial brightness profiles of the G ring at different longitudes relative to the arc's peak visible in (a). The profiles through the arc (gray) and elsewhere (black) are essentially identical beyond 168 000 km, whereas the arc has a sharp peak at 167 500 km. The absorption feature's radial width is comparable to that of the visible arc. While the absorption is radially displaced from the arc, this may be explained by larger-scale magnetospheric processes or through deformations in the magnetic field by the arc. (From Horanyi *et al.* (2009), after Hedman *et al.* (2007a); reprinted with permission from *Science*.)

Encke Gap

The Encke Gap is a 320-km-wide opening in Saturn's A ring that contains the small moon Pan and an array of dusty features composed of particles smaller than 100 μm. It includes three narrow ringlets that contain a series of bright clumps that do not follow the predicted orbits of isolated ring particles. Their eccentricities are induced by solar radiation pressure, and their positions are further perturbed by the combination of drag forces, gravitational perturbations, and collisions among the ring particles. Cassini even sees clumps merge and split. (See Figure 16.9.)

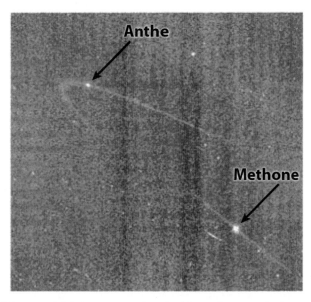

Figure 16.8 Image of debris arcs associated with small moons Anthe and Methone. (From Hedman *et al.* (2009): courtesy of NASA/JPL/Space Science Institute.)

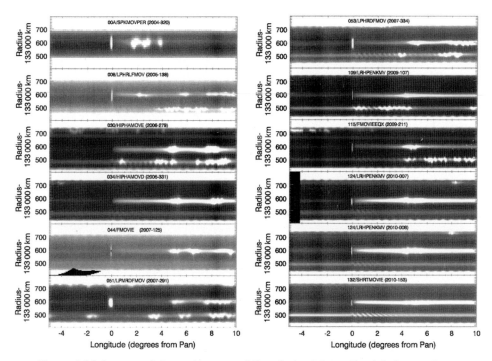

Figure 16.9 Images of the region around Pan derived from Cassini observations. Note the waves in the inner ringlet generated by Pan's gravitational perturbations. (From Hedman *et al.* (2013); reprinted with permission from Elsevier.)

E ring

The E ring is one of the broadest rings in the solar system, enveloping the icy satellites Mimas, Enceladus, Tethys, Dione, Rhea, and Titan. It has also been directly sampled by the Cosmic Dust Analyzer (CDA) experiment on Cassini. The small dust particles are dominantly water-ice grains, which we know are erupted from volcanoes on Enceladus, from the spectacular Cassini images of these eruptions. Cassini has measured these particles both at their source and out beyond the moon Titan at a distance of $20\,R_s$. Jürgen Schmidt and his colleagues suggest that particles in the volcanic plume condense from water vapor inside fractures in the moon's surface (the famous locations nicknamed "Tiger Stripes"). The ice grains are blown away from the moon by the escaping gas into the vacuum of space, where the ice particles (depending on their speed) can escape Enceladus' gravity and, if fortunate, avoid recollision with the moon and orbit Saturn to form the ring. Plasma and other forces change their orbits to spread the ring. Thus, the volcanic plumes are the ring source, completely different from the faint rings around other small moons, which are created by bombardment and collisions.

Spitale and Porco (2007) identify at least eight jets within the plume that contribute to the ring. Cassini remote sensing identifies these jet locations as the warmest spots on the Enceladus surface, and all located along the tiger stripes near the Enceladus south pole. The sizes of particles in the E ring are similar to those seen in the plume, about 1–5 μm in radius. The E-ring grains are composed mostly of water, but some are rich in sodium, perhaps derived from a salty sea below the Enceladus surface (Postberg *et al.*, 2011; Hansen *et al.*, 2011). The evaporation of this liquid water reservoir creates both the plumes and jets, and consequently Saturn's E ring.

Spokes

Spokes are intermittent, approximately radial markings on Saturn's B ring. First recognized by Voyager, they are thought to arise from small particles removed from larger ring particles by electrostatic repulsion. Voyager showed that the spokes are more common near dawn, when they seem to form, developing within minutes, becoming more intense in a few hours, and then later disappearing. While forming, they spread out, roughly following the Keplerian orbital motion. Their periodic behavior suggested they were triggered by Saturn's magnetic field.

Contrary to predictions, Cassini did not observe spokes until September 2005. This shows that the spoke variability is not just an observational artifact, but truly a seasonal phenomenon. Surprisingly, spokes are actually absent for long periods.

Figure 16.10 Top: Spokes in the B ring as seen by Voyager 2 (W. H. Smith *et al.*, 1981; B. A. Smith *et al.*, 1982). The left image was captured in backscattered light before closest encounter, with the spokes appearing as dark radial features across the ring's center. The right image was taken in forward-scattered light after the spacecraft crossed the ring plane and was looking back towards the Sun; the spokes now occur as bright markings. Typical dimensions of these spokes are 10 000 km in length and 2000 km in width. The nature of the changing brightness indicates that spokes consist of small grains (with radii < 1 μm), that is, that are comparable to the wavelength of visible light. At the time these images were taken, the rings' opening angle to the Sun was $B' = 8°$. Bottom: The initial spoke observations taken by Cassini on September 5, 2005 ($B' = 20.4°$), over a span of 27 min. These faint and narrow spokes were seen from the unilluminated side of the B ring. These spokes are approximately 3500 km long and approximately 100 km wide, much smaller than the average spokes seen by Voyager. These images were taken with a resolution of 17 km per pixel at a phase angle of 145° when Cassini was 13.5° above the unlit side of the rings as the spokes were about to enter Saturn's shadow. (From Mitchell *et al.* (2006); reprinted with permission from *Science*.)

Their disappearance may be caused by increased plasma density near the ring from sunlight producing photoelectrons and also sputtering neutral atoms from the ring particles. Near equinox, the sunlight is less. A similar lack of sunlight in Saturn's shadow may favor the production of spokes in the shadowed rings. (See Figure 16.10.)

Electric fields elevate the small dust particles that comprise the spokes, and afterwards they continue to orbit Saturn, visible until they are recollected by a larger ring particle. Most likely, spokes are triggered by meteoroid impacts onto the rings (Goetz and Morfill, 1983). The impact produces an expanding plasma cloud, which then cools as it moves radially away from Saturn. An alternative explanation is that lightning from Saturn thunderstorms produces a burst of electrons which travel along Saturn's magnetic field lines to strike the rings (Jones *et al.*, 2006; Horanyi *et al.*, 2010).

When Saturn passed through its equinox in 2009, spokes became more frequent and obvious: spoke activity matched the activity observed by Voyager a Saturnian year earlier (see Figure 16.10). Many other characteristics are similar as well (D'Aversa *et al.*, 2011; Mitchell *et al.*, 2013).

16.6 Summary

Tiny dust forms a highly visible but very small mass fraction of planetary rings. The dynamics of this dust is more complicated due to electromagnetic and radiation effects on the small particles in addition to gravitation. The short lifetimes mean that dust is continually created by micrometeoroid bombardment of small parent moons and by collisions between the larger ring bodies. Some dust can be captured by the magnetosphere or held by Jupiter's gravity after escaping a moon. This forms an exceedingly tenuous ring around Jupiter that extends to more than $50\,R_J$.

The remote observations of the main Jovian rings and the dust clouds around the Galilean satellites by the Galileo spacecraft provide a spectacular confirmation that dusty rings are probably dominated by impact ejecta and are quickly evolving.

The Cassini results show dust rings sculpted by orbital resonances; spokes triggered by Saturn's magnetosphere rotation or by lightning; drag forces that spread the ring arcs in longitude and away from Saturn. The inner D ring contains material that may be spiraling inward to Saturn as suggested for the Uranian rings by Colwell and Esposito (1990a). Images and remote spectroscopy from Cassini combine with direct measurements to give an excellent comprehensive picture of Saturn's diffuse rings. These show many parallels with the less studied rings of Jupiter, Uranus, and Neptune.

17

Concluding remarks

Planetary rings, those enduring symbols of space, exist around all the giant planets. Perhaps, once, every planet has had a ring. Earth once possessed a ring of debris which coalesced to form our own Moon. It is likely that planets around other stars will also have rings. (Some slight evidence from exoplanet studies supports this expectation.) In our solar system, each ring system is unique: this can be explained as the results of the stochastic events that create them from small nearby satellites. We ring scientists have built idealized models of ring physics which explain the major processes we observe. The apparent youth of the rings is a result of recent events that create them, like the shattering of a small moon. But that moon itself may be formed of constituents of former rings: thus nature recycles the basic raw material to renew rings. Rings are our best local laboratory for studying phenomena in flattened disks: many processes occurring now in rings resemble those in a protoplanetary disk where planets are forming, particularly interactions between the disk and embedded objects.

The possibility of rings surrounding the planets was not imagined by the ancients. Galileo's observations of Saturn's were a puzzle waiting explanation for decades. The interplay between observations and theory is a key aspect of our increasing understanding of planetary rings. With the new space observations of the last decades, the study of rings has been a dynamic, evolving enterprise, like the rings themselves.

From the first, the need to explain what we observe has driven new studies and expanded our understanding. Key examples are the close interactions of rings and moons, and the predominant role of stochastic events. Planetary rings, once curiosities restricted to the planet Saturn, are now the subject of detailed astrophysical and geophysical investigations.

The various planetary rings in our own solar system are each formed of immense numbers of small particles. The ring-particle composition resembles that of the nearby moons. We see broad and narrow rings; eccentric, inclined, and partial rings; empty gaps, and rings with moons and ringlets. Moons perturb rings, but also provide the sources and sinks for ring material. This process can be most clearly seen in Saturn's F ring.

Individual ring particles aggregate into loosely consolidated rubble piles. These grow and are disrupted by tides, collisions, and impacts. The larger ring particles may have a layer of smaller ring particles on their surfaces: these regoliths resemble the surfaces of icy moons. In addition to aggregation and disaggregation of the ring aggregates, the regoliths may be captured and lost, depending on their collisional environment. Cassini's images and other observations show that the nearby moons spur the creation of aggregates as seen in the F ring and the outer edges of Saturn's A and B rings. Such clumps were clearly seen at the time of equinox, when the Sun set on the rings. We can formally write the equations of ring physics, but the solutions are not easy. Simplifications help, but then we must wrestle with the fidelity of our hypotheses. Both hydrodynamic and kinetic approaches have proved useful.

Rings are confined by shepherding satellites: the ring edge forms, or the ring is truncated, at a resonance. At this radial location, the ratio of the orbital periods of the moon and ring particle is a ratio of small integers. The actions of gravity in this situation surprisingly result in a net repulsion. Many shepherding satellites have been found, although it is somewhat worrisome that more have *not* been found. The effects of small moons induce a pendulum-like motion of the ring particles near resonance. We find gaps, sharp edges, waves, and longitudinal and radial confinement at these locations. Moons in a gap can cause scalloped edges of nearby rings through edge wakes.

Rings show fine-scale structure, occasional dark lanes called spokes, bending and density waves. Some are more dusty than others, others have transient clumps. The F ring of Saturn appears braided. The dust in all ring systems must be continually replenished.

N-body simulations attempt to treat exactly the behavior of a very large sample of ring particles. These simulations show both the equilibrium and instabilities of ring systems. Some outputs from these simulations resemble unusual individual structures seen in real rings.

A stochastic approach to modeling planetary rings emphasizes the random events in ring history. Ring dynamics may be dominated by a limited number of the most massive bodies in the system. This leads to an unpredictable evolution: the history of a planetary ring depends on the outcome of a few rare random events. In this way, ring evolution resembles biological evolution. The actual histories are not close to the ensemble averages. The stochastic models therefore capture essential aspects of our planetary system.

The short lifetimes of small moons near giant planets mean they are rapidly destroyed. This implies a close relation between the rings and nearby moons: disrupted moons are the source for planetary rings. After a disruption, the largest fragments can act as shepherds, confining the remaining ring material radially and azimuthally. The entire ring–moon system may evolve as a unit. Satellites excite velocities in the rings, which, in turn, determine collisional outcomes, the size distribution, and produce dust. They can also cause new aggregates to form by driving the ring particles to converge where streamlines cross.

The balance between fragmentation and accretion determines the particle size distribution. A good test case is provided by long-term observations of Saturn's F ring by the Cassini orbiter. Repeated disruptions and aggregations can recycle the ring material many times, perhaps indefinitely extending the ring lifetime. If the rings are massive enough, their recycling could keep the rings young for the age of the solar system.

Neptune's puzzling rings still are incompletely explained. Competing theories can not be tested without close-up observations. Clearly, random events have played a major role in the history of Neptune's ring arcs. A large problem is to explain how these arcs might survive for millions of years, not to mention the age of the solar system.

Jupiter's rings are seen from Galileo observations to be closely associated with the nearby small moons. Each of the gossamer rings is derived from a small moon on an inclined orbit. The short lifetimes of the tiny ring particles require continuing replenishment from the ring-moons. Cassini's distant observations of Jupiter on the way to Saturn show that the particles in the main ring are non-spherical. A natural explanation for the particle size distribution that matches the Galileo data is that the Jovian main ring was created by the break-up of a progenitor body in the very recent past, perhaps only 1000 years ago.

Tiny dust particles form a highly visible, but very small mass, fraction of planetary rings. Dust dynamics is complicated by electromagnetic and radiation effects. Dust created by meteoroid bombardment is captured by Jupiter's gravity to form an exceedingly tenuous ring. A similar broad dust ring of Saturn probably originates from the retrograde moon Phoebe. Dust from Phoebe may spiral in to coat the leading hemisphere of Iapetus.

The results from Cassini provide a spectacular confirmation of the variety of phenomena in Saturn's rings and clues about the rings' history. Long-term investigations have now begun, to define the composition, structure, and dynamics of Saturn's rings. Further observations have followed the new discoveries. This has been coupled with renewed theoretical studies, which can partly explain our new observations. Before the Cassini mission ends, it will accurately measure the total mass of the rings. If Saturn's rings are truly ancient and continually renewed, they must be more massive (particularly ring B) than expected from the Voyager analysis of density waves. I favor a picture of Saturn's rings that is ancient and driven by stochastic events. Ring particles can aggregate, and later be broken apart to form new rings. Even if this hypothesis is still controversial, I can safely predict at least one thing. We will have continued surprises in the dynamic, evolving study of planetary rings!

17.1 The big questions

In the study of planetary rings, as in many areas of science, every new discovery leads to new questions. Our recent progress in understanding their physics leads naturally to several major open questions about planetary rings, as follows.

Origin and evolution

Just how old are the rings, where did they come from, and what is their future? Can we solidify the genetic relation between moons and rings? What is the contribution of external factors like meteoroid bombardment to ring history? What role does recycling of ring material play in ring history? Did Earth ever possess a ring, perhaps as an intermediate stage in forming our Moon? Why are Saturn's rings so much bigger than those around the other giant planets?

Ring make-up

What are the chemical composition, size distribution, and shape of the ring particles? How do these aspects combine to give the rings' appearance and other characteristics visible at a distance? Can we have a close-up view of individual ring particles?

Origins of planets

Obviously, the rings preserve some part of the history of their own planet and its system of satellites. How can we read this information? Will detailed study of ring dynamics clarify the processes that occur when a disk around a star coalesces to form planets? The rings we see are cold, with frequent collisions and large opacity; they exist in the region where the tides from the primary act to retard accretion (the Roche zone). These aspects are all different from the planet-forming environment. Despite these limitations, the rings show the effects of embedded large bodies, chaotic interactions, and azimuthally non-symmetric effects like density waves, bending waves, and warps. These phenomena are important in forming planets.

Furthermore, our experience is that answering the questions raised by explaining the real phenomena in planetary rings has stretched the existing theoretical and numerical methods. These improved methods have then been applied to other disk systems. This application of ring studies to understanding how planets form may be the largest payoff of ring research.

Cassini's final plunge

To answer the above questions, no single finding or measurement will suffice. In-depth, multi-instrument observations and linked theoretical developments are required. A particular opportunity is now provided by the joint NASA–ESA Cassini mission. Cassini has been investigating Saturn's rings since 2004; it observed the spectacular equinox which allowed us to see large objects in the rings by the long shadows they cast; it will eventually plunge inside the rings and measure their total mass.

Afterword
The Cassini Hall of Fame

The Cassini mission team has selected and collated their favorite images from the Cassini mission to Saturn and Titan, available on the Cassini Hall of Fame website at http://saturn.jpl.nasa.gov/photos/halloffame/. Some of these images unveil overwhelming beauty or show tricks of light or surprising oddities. Many include spectacular images of Saturn's rings. Some of the inspiration for the second edition of this book came from the unique and beautiful images taken by the Cassini mission around that rare moment of Saturn's equinox in 2009. Also see Figures 7.4, 7.7 and 7.14 in the text as these are also some of the best Hall of Fame images taken by Cassini: don't miss them. As an afterword to this book, I offer brief explanations for a selection of these images with a reference to where the observed phenomena are discussed in my text. The captions are taken from the Cassini website. These images confirm my personal view that planetary rings are both the most scientifically intriguing and beautiful objects in the universe.

Figure A-1, **Cassini Hall of Fame image #3**. "Saturn's A Ring From the Inside Out." The best view of Saturn's rings in the ultraviolet indicates there is more ice toward the outer part of the rings than in the inner part, hinting at the origins of the rings and their evolution. Images taken during the Cassini spacecraft's orbital insertion on June 30, 2004, show compositional variation in the A, B, and C rings. From the inside out, the Cassini Division in faint red at left is followed by the A ring in its entirety. See Section 4.5 for the composition of Saturn's rings. (PIA05075; courtesy of NASA/JPL/University of Colorado.) For color version, see Plates section.

Figure A-2, **Cassini Hall of Fame image #7**. "The Face of Beauty." Few sights in the solar system are more strikingly beautiful than softly hued Saturn embraced by the shadows of its stately rings. Despite Cassini's revelations, Saturn remains a world of mystery. The shadows of Saturn's rings are beautiful, but not explained further in this book. (PIA07772; courtesy of NASA/JPL/Space Science Institute.) For color version, see Plates section.

Figure A-3, **Cassini Hall of Fame image #12**. "Saturn's C and B Rings From the Inside Out." Images taken during the Cassini spacecraft's orbital insertion on June 30, 2004, show definite compositional variation within the rings. This image shows, from left to right, the outer portion of the C ring and inner portion of the B ring. The B ring begins a little more than halfway across the image. The general pattern is from "dirty" particles nearer Saturn to cleaner ice particles shown in turquoise in the outer parts of the rings. Ring composition is discussed in Section 4.5. (PIA05076; courtesy of NASA/JPL/University of Colorado.) For color version, see Plates section.

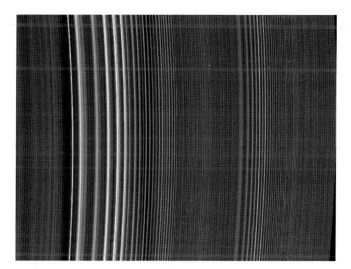

Figure A-4, **Cassini Hall of Fame image #17**. "Catching Saturn's Ring Waves." This false color image of two density waves in Saturn's A ring was made from the stellar occultation observed by Cassini's ultraviolet imaging spectrograph when the spacecraft was 6.3 million km (4 million miles) from Saturn. For the theory of density waves, see Section 6.4. (PIA06994; courtesy of NASA/JPL/University of Colorado at Boulder.) For color version, see Plates section.

Figure A-5, **Cassini Hall of Fame image #22**. "Waves and Small Particles in Ring A." Specially designed Cassini orbits place Earth and Cassini on opposite sides of Saturn's rings, a geometry known as radio occultation. Cassini conducted the first radio occultation observation of Saturn's rings on May 3, 2005. Three simultaneous radio signals at wavelengths of 0.94, 3.6, and 13 cm (Ka-, X-, and S-bands) were sent from Cassini through the rings to Earth. The observed change of each signal as Cassini moved behind the rings provided a profile of the distribution of ring material and an optical depth profile. This computer-generated image provides a color rendition of the optical depth in Saturn's rings. For the definition of optical depth, see Section 15.1. (PIA07960; courtesy of NASA/JPL.) For color version, see Plates section.

Figure A-6, **Cassini Hall of Fame image #27**. "Moon-Made Rings." This view, acquired with the sun almost directly behind Saturn, reveals a previously unknown faint ring of material coincident with the orbit of the newly discovered small moon Pallene and a new diffuse ring that is coincident with the orbits of Saturn's moons Janus and Epimetheus. See Section 16.5. (PIA08328; courtesy of NASA/JPL/Space Science Institute.)

Figure A-7, **Cassini Hall of Fame image #31**. "On the Final Frontier." Saturn sits nested in its rings of ice as Cassini once again plunges toward the graceful giant. This natural color mosaic was acquired by the Cassini spacecraft as it soared 39° above the unilluminated side of the rings. Saturn's shadow falls on the rings. (PIA08388; courtesy of NASA/JPL/Space Science Institute.) For color version, see Plates section.

Figure A-8, **Cassini Hall of Fame image #32**. "Saturn's Rings in the Infrared." This mosaic of Saturn's rings was acquired by Cassini's visual and infrared mapping spectrometer instrument on September 15, 2006, while the spacecraft was in the shadow of the planet looking back towards the rings from a distance of 2.16 million km (1.34 million miles). Data at wavelengths of $1.0\,\mu m$, $1.75\,\mu m$, and $3.6\,\mu m$ were combined in the blue, green, and red channels to make the pseudo-color image shown here. This pseudo-color image show the rings as would be seen in infrared. (PIA01940; courtesy of NASA/JPL/University of Arizona.) For color version, see Plates section.

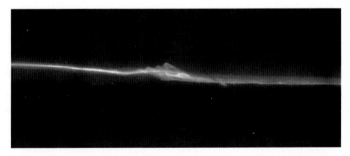

Figure A-9, **Cassini Hall of Fame image #37**. "F Ring Dynamism." The Cassini spacecraft has revealed a never-before-seen level of detail in Saturn's F ring, including evidence for the perturbing effect of small moonlets orbiting in or close to the ring's bright core. For some time, scientists have suspected the presence of tiny moonlets that orbit Saturn in association with the clumpy ring. As the small satellites move close to the F-ring core they leave a gravitational signature. In some cases they can draw out material in the form of a "streamer" – a miniature version of the interaction Cassini has witnessed between Prometheus and the F-ring material. The dynamics of this interaction are the same, but the scale is different. Evidence for small moonlets in the F ring: see Section 11.5. (PIA08290; courtesy of NASA/JPL/Space Science Institute.)

Figure A-10, **Cassini Hall of Fame image #42**. "Mapping Clumps in Saturn's Rings." This false-color image of Saturn's main rings was made by combining data from multiple star occultations using the Cassini ultraviolet imaging spectrograph. During occultations, scientists observe the brightness of a star as the rings pass in front of the star. This provides a measurement of the amount of ring material between the spacecraft and the star. Self-gravity wakes are discussed in Sections 7.6 and 8.4. (PIA09210; courtesy of NASA/JPL/University of Colorado.)

Figure A-11, **Cassini Hall of Fame image #46**. "Secretic Rings." Pan is seen in the Encke Gap, halfway between the center and the right. See also Figure 7.16. (PIA09758; courtesy of NASA/JPL/Space Science Institute.)

Figure A-12, **Cassini Hall of Fame image #50**. "Rings Aglow." Saturn's softly glowing rings shine in scattered sunlight. The B ring presents a remarkable difference in brightness between the near and far arms (bottom and top of the image, respectively). The strong variation in brightness could be due to the presence of wake-like features in the B ring. The azimuthal brightness asymmetry is discussed in Section 7.1. See Figure 7.9. (PIA09860; courtesy of NASA/JPL/Space Science Institute.) For color version, see Plates section.

Figure A-13, **Cassini Hall of Fame image #54**. "Scattered Sunshine." Saturn's icy rings shine in scattered sunlight in this view, which looks toward the unilluminated northern side of the rings from about 15° above the ring plane. The Sun currently illuminates the rings from the south. Some of the sunlight not reflected from the rings' southern face is scattered through the countless particles, setting the rings aglow. Diffuse transmission and reflection by the rings is discussed in Section 15.1. (PIA10446; courtesy of NASA/JPL/Space Science Institute.) For color version, see Plates section.

Figure A-14, **Cassini Hall of Fame image #58**. "With Ghostly Spokes." Saturn's moon Pandora shares the stage with ghostly B-ring spokes in this Cassini spacecraft scene. Pandora (81 km across) is on the left. The spokes are the radial markings visible on the right of the image. See Sections 7.3 and 16.5. (PIA12547; courtesy of NASA/JPL/Space Science Institute.)

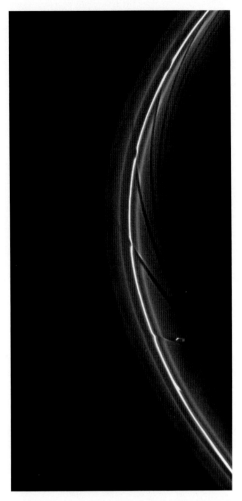

Figure A-15, **Cassini Hall of Fame image #62**. "Fleeing the Scene." Saturn's moon Prometheus, having perturbed the planet's thin F ring, continues in its orbit. The gravity of potato-shaped Prometheus (86 km across) periodically creates streamer-channels in the F ring, and the moon's handiwork can be seen in the dark channels here. See Section 11.2. (PIA12684; courtesy of NASA/JPL/Space Science Institute.)

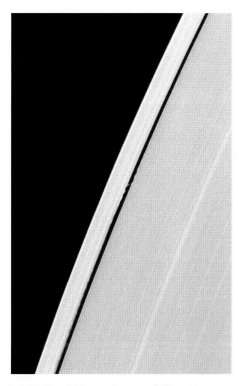

Figure A-16, **Cassini Hall of Fame image #68**. "Disturbances by Daphnis." Daphnis cruises through the Keeler Gap, raising edge waves in the ring material as it passes. As is characteristic of waves raised by a moon on the edges of a very narrow gap like Keeler, the wave begins as a coherent form near Daphnis and becomes less so with increasing orbital distance from the moon. See Section 6.3. (PIA08882; courtesy of NASA/JPL/Space Science Institute.)

Glossary

accretion (adj: accretional) the process of growth of small particles to larger sizes by collisions that result in their sticking together. Also referred to as *coagulation*.

accretion disk the flat disk surrounding a planet or star where the particles are coagulating into planets and moons as the result of their collisions.

albedo the fraction of light reflected by a body. Astronomers distinguish the plane albedo (light reflected by a small surface element), spherical albedo (reflected by the entire sphere illuminated by the Sun), geometric albedo (fraction reflected back to an observer at the exact moment of opposition), and bond albedo (reflectivity averaged over all wavelengths).

apoapse furthest point in an orbit from the central body. Apochrone: furthest point from Saturn. See *periapse*.

apsides plural of *apse*. Line of apsides: the line connecting the closest and furthest points of an orbit.

arc incomplete ring. Also *ring arc*.

axisymmetric symmetric about some line or axis. Possessing cylindrical symmetry. A common, but often incorrect, assumption about planetary rings.

azimuth (adj: azimuthal) the angle measuring longitude. Having azimuthal symmetry: see *axisymmetric*.

azimuthal brightness variation (or asymmetry) The brightness of a ring varies with the viewing direction for the A and B rings of Saturn. Explained by *self-gravity wakes*.

azimuthal confinement confinement of the ring particles to some small range of longitude.

ballistic transport moving material by *meteoroid* bombardment and subsequent ejection and recapture of the fragments. Between ejection and recapture, the ring material moves on a ballistic trajectory.

bending wave a spiral wave excited at a location where the ring particle's orbital period is in *resonance* with the orbital period of an inclined moon. The particle's inclination is enhanced, and the local ring plane is tilted. This results in a corrugation of the ring. The bending wave propagates away from the perturbing moon. See also *density wave*.

Boltzmann equation equation describing the rate of change of a distribution of particles resulting from the various forces and collisions acting on them.

braids, braided pattern in which ring segments or *strands* apparently cross each other.

Cartesian deriving from or related to the ideas of René Descartes. He hypothesized that celestial motions involved swirling material (Cartesian vortices) and invented plane algebraic geometry (Cartesian coordinates and axes).

coagulation sticking together of small bodies to form larger ones. This stage of planetary formation follows the condensation of dust grains. In planetary rings, coagulation competes with fragmentation.

coagulation equation a differential equation that gives the rate of *accretion* in terms of the collisional properties of the particles in the system.

collisional cascade the process where repeated collisions break a system of bodies into successively smaller fragments.

collision integral the term on the right-hand side of the *Boltzmann equation* that determines the total effect of collisions on changing the particles' distribution.

coplanar lying in the same plane.

corotation orbital period that is the same as the planet's rotation period. For Earth, this point is called "geosynchronous," and is useful for communication satellites.

corotation inclination resonance (CIR) a location in a ring where the combination of the orbital motion and the vertical oscillation frequency of a ring particle resonates with the forcing frequency of a moon.

corotation eccentricity resonance (CER) a location in a ring where the combination of the orbital motion and the epicyclic motion of a ring particle resonates with the forcing frequency of a moon.

cosmic recycling the hypothesis that the material in planetary rings is used multiple times, cycling between rings of material and partly consolidated *rubble piles*.

critical density the minimum density of an orbiting object that, at a given distance from its planet, is able to hold itself together by self-gravity.

density wave a spiral wave excited at a location where the ring particle's orbital period is in *resonance* with the orbital period of a moon. The particle's eccentricity is enhanced at the resonance, and the local density of ring particles increases in the crests of the waves, and decreases in the wave troughs. The wave propagates toward the perturbing moon. See also *bending wave*.

differential rotation in a planetary ring, faster orbiting of the particles nearer the planet, due to Kepler's third law. This results in *Kepler shear*.

diffusional instability a possible instability of a flat disk, causing a homogeneous disk to divide into discrete regions of high density with low velocity separated by regions of low density with higher velocity.

dipole having just two poles, for example, north and south. More complicated distributions need to be expressed in a multipole expansion, where the higher, more complicated terms are quadrupole, octopole, etc. Far from a planet, the dipole alone provides a good approximation for the magnetic field.

discretized for a continuous process, reduced to a set of values at particular times or locations.

drunkard's walk an alternative term for the *random walk*, the path followed by a particle that randomly moves either left or right.

dynamic ephemeral bodies (DEBs) a model for ring particles composed of many smaller elements that continually collect and then disperse. The unconsolidated bodies are under-dense and short-lived.

epicycle the small circle (actually an ellipse) that a particle on an eccentric orbit appears to make around a point following a perfectly circular orbit. This approximation is valid for low-eccentricity orbits, like those of particles in planetary rings.

epicyclic frequency the rate at which a particle on an eccentric orbit executes its apparent *epicycle* motion. For orbits around a spherical planet, this frequency is equal to the orbital frequency, and the orbit is therefore closed. For an oblate planet, the epicyclic frequency is slightly less than the orbital frequency, the orbit is not a closed curve, and thus the orbit precesses.

equipartition the condition where, in a system of particles, all degrees of freedom share the energy equally. For a system of ring particles with a range of masses, this requires that the smaller particles have a higher average velocity.

exosphere the outer part of a planet's atmosphere, where the atoms and molecules are escaping the planet's gravity.

forced, forcing see *gravitational perturbation*.

gas drag the frictional force experienced by ring particles colliding with the gas molecules within the uppermost atmosphere of a planet.

gossamer very light and sheer, describing the ethereal rings of Jupiter lying outside its main rings.

gravitational perturbation alteration of the orbital motion of ring particles by the gravitational pull of satellites, however small. If this is periodic, the ring particle feels the pull at the "forcing frequency;" an initially circular orbit may attain a "forced eccentricity."

harmonic oscillator a physical system, like a mass on a spring, that oscillates with a fixed period. This may be a good model for the dynamics of ring particles.

inelastic not conserving energy. The collision between two ring particles is generally inelastic.

isotropic the same in all directions; no dependence of directionality. Anisotropic: not the same in all directions. Noun: anisotropy.

jet a feature in Saturn's F ring that appears as a small streak, evidence of a collision in the ring. See also *mini-jet*.

Jovian pertaining to the planet Jupiter.

Kepler developer of laws of planetary motion, "Kepler's laws." Particles obeying Kepler's laws travel in an ellipse: this is called "Keplerian motion." Particles further from the planet travel more slowly; the decline in orbital speed gives *Kepler shear*.

Kepler shear the gradient in which particles orbiting a central body have decreasing velocity further from it.

kinetic theory a description of a gas as a large number of small particles in constant random motion.

kittens transient clumps in Saturn's F ring.

libration, librate the motion back and forth around an equilibrium point. The libration of the Moon causes more than just one-half of the Moon to be visible to earthly observers.

Lindblad resonance a location in the rings where the epicyclic motion of a ring particle resonates with the forcing frequency from a moon's gravity.

Liouville equation the equation for the evolution of a system in a *phase space* of position and velocity.

lossy losing energy, *inelastic*.

magnetosphere the region surrounding a planet that is controlled by its own magnetic field.

Markov process a stochastic process whose evolution depends only on its present state (the *Markov property*).

Markov property the property of a random process that its probabilities for future evolution depend only on its present state, and not on its past history

mean free path the average distance a particle travels between collisions.

meteoroid a small particle in space. The frequency of meteoroids is given by the "meteoroid (or meteoritic) flux."

microgravity gravitational acceleration much smaller than at the Earth's surface.

micrometeoroid a small *meteoroid*, generally less than 1 mm in size.

microstructure small-scale structure observed in Saturn's rings, generally less than 1000 km in size. This structure extends down to scales as small as 15 km across.

mini-jet a small trail, 40–80 km long, seen in Saturn's F ring that is the likely result from a collision there.

moment equation an equation derived by multiplying by a power of the velocity and integrating. For the *Boltzmann equation*, this yields energy and momentum equations.

monolayer a layer of ring particles only one particle thick. Alternatively, we have a *multilayer*: a layer of particles many particles thick.

Monte Carlo method a computer method to simulate the probabilistic evolution of a physical system by drawing numbers randomly to determine successive outcomes.

moonlet a small moon, generally with radius less than 10 km. The distinction between the smallest "moonlet" and the largest "ring particle" may not always be clear.

moonlet belt a ring composed of *moonlets*.

mutual Hill sphere a sphere with radius R_H in Hill's approximation, as applied to two particles. Within this distance, the gravitational acceleration between the two particles typically dominates over the difference in force from the planet's gravity.

***N*-body simulation** a computer model where a large number of individual bodies are followed to simulate a larger population, say the particles in a planetary ring.

node the point at which an inclined orbit crosses the equatorial plane.

noise random fluctuations that interfere with a measurement. It is said that "One man's data is another man's noise."

oblate not perfectly spherical, but flattened at the poles. All the giant planets are oblate.

occultation blockage by rings of a source of radiation, such as a star (stellar occultation) or the spacecraft radio transmitter (radio occultation).

optical depth the ability of a ring to block light. This is directly measured by an *occultation*.

orthonormal mutually perpendicular axes, having unit length.

overstability a pulsational instability that spontaneously forms waves that travel in the radial direction in Saturn's rings.

parent bodies the source bodies for particles in planetary rings.

periapse the point in an orbit of closest approach to the primary. Perichrone: closest approach to the planet Saturn.

perturbation, perturbed see *gravitational perturbation*.

phase angle The angle between the Sun and the viewer, as seen from the illuminated body.

phase function the function that describes the angular redistribution of light after a scattering event.

phase space in mathematical physics, a multi-dimensional space where every possible state of a system is represented by a unique point in the phase space.

photometry measurement of light.

photopolarimeter an astronomical instrument for measuring the brightness and polarization of light.

planetesimals the small bodies that grew to become planets.

plasma a gas of ionized particles.

plasma drag the frictional force a ring particle feels as it orbits through ionized particles (*plasma*) surrounding a planet.

point-mass planet the approximation that all the mass of a planet is concentrated at a single point.

power law a functional form, where the dependent variable is found as the power of the independent variable. For example, the area of a circle varies as the second power (square) of the radius.

power-law index the numerical value of the power in a *power-law* distribution.

power spectrum the distribution of total power versus frequency.

Poynting–Robertson drag (effect) a relativistic effect due to absorption and readmission of light that acts to slow a particle orbiting a luminous object.

precession the behavior of a particle orbiting an oblate planet to advance its closest approach point: for example, "precession of *periapse*."

pressure tensor a matrix representing the stresses on a small volume in a planetary ring.

probability density a function defined so that the integral between two limits gives the probability of finding the system between those limits.

prograde forward. For planets, the prograde direction is counter-clockwise as viewed from the north. Opposite: *retrograde*.

propeller the S-shaped structure in a ring created by a small embedded moon that is too small to clear out a complete gap.

protoplanets bodies on their way to becoming planets.

proximal nearest, meaning the Cassini orbits at the end of its mission that come closest to, and eventually strike, the planet.

quadrupole having four poles. A term in the "multipole" expansion of a gravitational or other field. The coefficient in this expansion is the "quadrupole moment."

radiation belt trapped radiation, like the Earth's Van Allen belts.

radiative transfer the study of radiation transported in a scattering and/or absorbing medium.

random walk a mathematical model of random motion. A particle moves either right or left at regular intervals, the directions chosen randomly. Also called *drunkard's walk*.

regolith the soil of unconsolidated material on the surface of a ring particle.

regression moving backwards. For a particle orbiting an *oblate* planet, the *node* regresses.

resonance the occurrence of two periodic motions at a ratio of small numbers. For example, the Cassini Division is located where the ring particle orbital period is exactly one-half of the period of the moon, Mimas. This is termed the 2 : 1 resonance. This resonance causes a resonant *perturbation* of the ring particle's orbit.

retrograde backwards orbital or rotational motion opposite to that of the planets. From the north, this appears clockwise.

ring arc an incomplete ring.

ringlet a narrow ring feature.

ring–moon a moon near or within a ring.

ring precursors parent bodies for ring material: when a moon is destroyed, a ring is formed.

Roche limit the distance from a planet, inside of which a fluid body would be disrupted since the tidal stresses from the planet's gravity exceed the gravity holding the body together. See also *Roche zone*.

Roche zone the broad region surrounding a planet, and including the classical *Roche limit*, where gravitational attraction competes with tidal disruption.

rubble pile a model for ring particles, asteroids, and small moons that proposes they are not solid, but unconsolidated, essentially a "pile of rubble." See also *dynamic ephemeral bodies (DEBs)*.

Saturn electrostatic discharges (SED) bursts of radio-frequency noise that may originate in atmospheric electrical storms or in Saturn's rings.

Saturn kilometric radiation (SKR) long-wave radiation emitted from Saturn.

self-gravity wakes temporary elongated clumps held together by gravity.

semi-major axis half the major axis of an elliptical orbit.

separatrix the curve separating two distinctive types of behavior, for example circulating and librating orbits.

shepherds short for "shepherd moons." Moons whose gravity holds a ring in place.

spectroscopy analysis of the spectra of light.

spokes radial dark lanes seen in Saturn's B ring.

sputtering a process where atoms are ejected from a solid object struck by energetic particles.

stochastic an approach for modeling a physical system as subject to probabilistic laws and evolution.

stochastic process a system that evolves following probabilistic laws.

strands narrow components of a ring; a particular case is that of Saturn's F ring, which appeared to Voyager to be composed of four main strands.

stress tensor see *pressure tensor*.

tilt effect the variation of ring brightness with the seasonal variation in ring opening angle.

toroidal donut-shaped.

torque the product of force and lever arm distance.

transition matrix for a *Markov process*, the matrix of probabilities for changing from one state of the system to another.

transport coefficients coefficients in a transport equation, for example, viscosity, conductivity, etc.

Trojan asteroids asteroids that follow or lead Jupiter by 60°.

unbounded without an upper or lower bound.

vertical resonance a location in the rings where the vertical oscillation of a ring particle resonates with the forcing frequency of a moon's gravity.

viscosity the characteristic of a fluid to transfer momentum between fluid elements moving at different velocity through collisions. Also, the numerical value of the viscosity.

viscous instability a possible amplification of density differences in a planetary ring excited by viscous transfer of momentum in collisions. See also *diffusional instability*.

viscous transport the transfer of momentum and energy through particle collisions.

wake the wavy phenomenon left in a ring by a passing moon.

zonal harmonics the elements in a spherical harmonic expansion of a gravitational or other field.

References

Albers, N. and Spahn, F. (2006). The influence of particle adhesion on the stability of agglomerates in Saturn's rings. *Icarus*, **181**, 292–301.

Albers, N., Sremcevic, M., Colwell, J. E., and Esposito, L. W. (2012). Saturn's F ring as seen by Cassini UVIS: Kinematics and statistics. *Icarus*, **217**, 367–88.

Alexander, A. F. O'D. (1962). *The Planet Saturn: A History of Observation, Theory, and Discovery*. New York: Dover.

Araki, S. and Tremaine, S. D. (1986). The dynamics of dense particle disks. *Icarus*, **65**, 83–109.

Attree, N. O., Murray, C. D., Cooper, N. J., and Williams, G. A. (2012). Detection of low-velocity collisions in Saturn's F ring. *Astrophys. J. Lett.*, **755**, 2, L27.

Baillie, K., Colwell, J. E., Lissauer, J. J., Esposito, L. W., and Sremcevic M. (2011). Waves in Cassini UVIS stellar occultations 2. Waves in the C ring. *Icarus*, **216**(1), 292–308.

Baldi, P., Mazliak, L., and Priouret, P. (2002). *Martingales and Markov Chains: Solved Exercises and Elements of Theory*. Boca Raton, FL: Chapman & Hall/CRC Press, p. 92.

Barbara, J. M. and Esposito, L. W. (2002). Moonlet collisions and the effects of tidally modified accretion in Saturn's F ring. *Icarus*, **160**, 161–71.

Baum, W. A., Kreidl, T., Westphal, J. A. *et al.* (1981). Saturn's E-ring. I. CCD observations of March 1980. *Icarus*, **47**, 84–96.

Beurle, K., Murray, C. D., Williams, G. A., *et al.* (2010). Direct evidence for gravitational instability and moonlet formation in Saturn's rings, *Astrophys. J. Lett.*, **718**, L176–L180.

Borderies, N. P., Goldreich, P., and Tremaine, S. D. (1984). Unsolved problems in planetary rings dynamics. In *Planetary Rings*, ed. R. Greenberg and A. Brahic. Tucson, AZ: University of Arizona Press, pp. 713–736.

(1985). A granular flow model for dense planetary rings. *Icarus*, **63**, 406–20.

(1989). The formation of sharp edges in planetary rings by nearby satellites. *Icarus*, **80**, 344–60.

Bosh, A. S., Olkin, C. B., French, R. G., and Nicholson, P. D. (2002). Saturn's F ring: kinematics and particle sizes from stellar occultation studies. *Icarus*, **157**, 57–75.

Bradley, E. T., Colwell, J. E., Esposito, L. W., *et al.* (2010). Far ultraviolet spectral properties of Saturn's rings from Cassini UVIS. *Icarus*, **206**, 458–66.

Brahic, A. (1975). A numerical study of a gravitating system of colliding particles: applications to the dynamics of Saturn's rings and to the formation of the solar system. *Icarus*, **25**, 452–8.

(1977). Systems of colliding bodies in a gravitational field: numerical simulation of the standard model. *Astron. Astrophys.*, **54**, 895–907.

Bridges, F. G., Hatzes, A. P., and Lin, D. N. C. (1984). Structure, stability, and evolution of Saturn's rings. *Nature*, **309**, 333–5.

Brooks, S. (2003). Jupiter's ring system revisited: a deeper understanding from Galileo visible and infrared observations. Ph.D. thesis, University of Colorado, Boulder, CO.

Brooks, S., Esposito, L. W., Throop, H. B., and Showalter, M. R. (2002). Photometric analysis of Jupiter's main ring with Galileo SSI and NIMS imaging. Paper presented at the International Jupiter Conference, Boulder, CO.

(2004). The size distribution of Jupiter's main ring from Galileo imaging and spectroscopy. *Icarus*, **170**, 35–57.

Brophy, T. G. and Esposito, L. W. (1989). Simulation of collisional transport processes and the stability of planetary rings. *Icarus*, **78**, 181–205.

Brophy, T. G., Stewart, G. R., and Esposito, L. W. (1990). A phase-space fluid simulation of a two-component narrow planetary ring: particle size segregation, edge formation, and spreading rates. *Icarus*, **83**, 133–55.

Brophy, T. G., Esposito, L. W., Stewart, G. R., and Rosen, P. D. (1992). Numerical simulation of satellite-ring interactions: resonances and satellite-ring torques. *Icarus*, **100**, 412–33.

Brown, R. H., Baines, K. H., Bellucci, G., *et al.* (2006). Observations in the Saturn System during approach and orbital insertion, with Cassini's Visual and Infrared Mapping Spectrometer (VIMS). *Astron. Astrophys.*, **446**, 707–716.

Brush, S. G., Everitt, C. W. F., and Garber, E. (eds.) (1983). *Maxwell on Saturn's Rings: James Clerk Maxwell's Unpublished Manuscripts and Letters on the Stability of Saturn's Rings*. Cambridge, MA: MIT Press.

Burns, J. A. (1999). Planetary rings. In *The New Solar System*, ed. J. K. Beatty, C. C. Petersen, and A. Chaikin. Cambridge: Cambridge University Press, pp. 221–40.

Burns, J. A., Showalter, M. R., Cuzzi, J. N., and Pollack, J. B. (1980). Physical processes in Jupiter's ring: clues to its origin by Jove! *Icarus*, **44**, 339–60.

Burns, J. A., Showalter, M. R., and Morfill, G. E. (1984). The ethereal rings of Jupiter and Saturn. In *Planetary Rings*, ed. R. Greenberg and A. Brahic. Tucson, AZ: University of Arizona Press, pp. 200–72.

Burns, J. A., Showalter, M. R., Hamilton, D. P., *et al.* (1999). The formation of Jupiter's faint ring. *Science*, **284**, 1146–50.

Burns, J. A., Hamilton, D. P., and Showalter, M. R. (2001). Dusty rings and circumplanetary dust: observations and simple physics. In *Interplanetary Dust*, ed. E. Grün, B. A. S. Gustafson, S. F. Dermott, and H. Fechtig. Berlin: Springer-Verlag, pp. 641–725.

Burns, J. A., Simonelli, D. P., Showalter, M. R., *et al.* (2004). Jupiter's ring–moon system. In *Jupiter, The Planet, Satellites and Magnetosphere*, ed. F. Bagenal, T. Dowling, and W. McKinnon. Cambridge: Cambridge University Press.

Camichel, H. (1958). Mesures photométriques de Saturne et son anneau. *Ann. Astrophys.*, **21**, 231–42.

Canup, R. M. (2010). Origin of Saturn's rings and inner moons by mass removal from a lost Titan-sized satellite. *Nature*, **468**, 943–26.

Canup, R. M. and Esposito, L. W. (1995). Accretion in the Roche zone: Coexistence of rings and ringmoons. *Icarus*, **113**, 331–52.

Canup, R. M. and Esposito, L. W. (1997). Evolution of the G ring and the population of macroscopic ring particles. *Icarus*, **126**, 28–41.

Canup, R. M., Colwell, J. E., and Horanyi, M. (1993). Size distributions of satellite dust ejecta: effects of radiation pressure and planetary oblateness. *Icarus*, **105**, 363–9.

Chandrasekhar, S. (1969). *Ellipsoidal Figures of Equilibrium*. New Haven, CT: Yale University Press.

Charnoz S. (2009). Physical collisions of moonlets and clumps with the Saturn's F-ring core. *Icarus*, **201**, 191–7.

Charnoz, S., Porco, C. C., Deau, E., *et al.* (2005). Cassini discovers a kinematic spiral ring around Saturn. *Science*, **310**(5752), 1300–4.

Charnoz, S., Dones, L., Esposito, L. W., Estrada, P. R., and Hedman, M. M. (2009). Origin and evolution of Saturn's ring system. In *Saturn From Cassini–Huygens*, ed. M. Dougherty, L. W. Esposito, and S. Krimigis. Dordrecht: Springer-Verlag, pp. 537–75.

Chiang, E. I. and Goldreich, P. (2001). Apse alignment of narrow eccentric planetary rings. *Astrophys. J.*, **540**, 1084–90.

Clark, R. N., Carlson, R., Grundy, W., and Noll, K. (2013a) Observed ices in the Solar System. In *The Science of Solar System Ices*, ed. M Gudipati and J. Castillo-Rogez. Astrophysics and Space Science Library 356, New York: Springer, pp. 3–46.

Clark, R. N., Swayze, G. A., Carlson, R., Grundy, W., and Noll, K. (2013b). Spectroscopy from space. In *Spectroscopic Methods in Mineralogy and Material Sciences*, ed. G. Henderson, in press.

Colwell, J. E. (2003). Low velocity impacts into dust: results from the COLLIDE-2 microgravity experiment. *Icarus*, **164**, 188–96.

Colwell, J. E. and Esposito, L. W. (1990a). A numerical model of the Uranian dust rings. *Icarus*, **86**, 530–60.

(1990b). A model of dust production in the Neptune ring system. *Geophys. Res. Lett.*, **17**, 1741–4.

(1992). Origins of the rings of Uranus and Neptune. I. Statistics of satellite disruptions. *J. Geophys. Res.*, **97**(10), 227–41.

(1993). Origins of the rings of Uranus and Neptune. II. Initial distributions of disrupted satellite fragments. *J. Geophys. Res.*, **98**, 7387–401.

Colwell, J. E. and Taylor, M. (1999). Low velocity microgravity impact experiments into simulated regolith. *Icarus*, **138**, 241–9.

Colwell, J. E., Horanyi, M., and Grün, E. (1998). Capture of interplanetary and interstellar dust by the Jovian magnetosphere. *Science*, **280**, 88–91.

Colwell, J. E., Esposito, L. W., and Bundy, D. (2000). Fragmentation rates of small satellites in the outer solar system. *J. Geophys. Res.*, **105**, 17, 589–99.

Colwell, J. E., Esposito, L. W., and Sremcevic, M. (2006). Self-gravity wakes in Saturn's A ring measured by stellar occultations from Cassini. *Geophys. Res. Lett.*, **33**, L07201.

Colwell, J. E., Esposito, L. W., Sremcevic, M., Stewart, G. R. and McClintock, W. E. (2007). Self-gravity wakes and radial structure of Saturn's B ring. *Icarus*, **190**, 127–44.

Colwell, J. E., Cooney, J. H., Esposito, L. W., and Sremcevic, M. (2009a). Density waves in Cassini UVIS stellar occultations. I. The Cassini Division. *Icarus*, **200**, 574–80.

Colwell, J. E., Nicholson, P. D., Tiscareno, M. S., *et al.* (2009b). The structure of Saturn's rings. In *Saturn From Cassini–Huygens*, ed. M. Dougherty, L. W. Esposito, and S. Krimigis. Dordrecht: Springer-Verlag, pp. 375–412.

Cook, A. F. and Franklin, F. A. (1964). Rediscussion of Maxwell's Adams prize essay on the stability of Saturn's rings. *Astron. J.*, **69**, 173–200.

Crida, A. and Charnoz, S. (2010). Solar system: Recipe for making Saturn's rings. *Nature*, **468**, 903–5.

Cuzzi, J. N. (1985). The rings of Uranus: Not so thick, not so black. *Icarus*, **63**, 312–16.

(1995). Evolution of planetary ringmoon systems. *Earth, Moon, and Planets*, **67**, 179–208.

(1998). Ringside seat. *The Sciences*, **38**, 28–33.

Cuzzi, J. N. and Burns, J. A. (1988). Charged particle depletion surrounding Saturn's F ring: evidence for a moonlet belt? *Icarus*, **74**, 284–324.

Cuzzi, J. N. and Durisen, R. H. (1990). Bombardment of planetary rings by meteoroids: general formulation and effects of Oort cloud projectiles. *Icarus*, **84**, 467–501.

Cuzzi, J. N. and Estrada, P. R. (1998). Compositional evolution of Saturn's rings due to meteoroid bombardment. *Icarus*, **132**, 1–35.

Cuzzi, J. N., Lissauer, J. J., Esposito, L. W., *et al.* (1984). Saturn's rings: properties and processes. In *Planetary Rings*, ed. R. Greenberg and A. Brahic. Tucson, AZ: University of Arizona Press, pp. 73–199.

Cuzzi, J., Clark, R., Filacchione, G., *et al.* (2009). Ring particle composition and size distribution. In *Saturn From Cassini–Huygens*, ed. M. Dougherty, L. W. Esposito, and S. Krimigis. Dordrecht: Springer-Verlag, pp. 459–509.

Cuzzi, J. N., Burns, J. A., Charnoz, S., *et al.* (2010). An evolving view of Saturn's dynamic rings. *Science*, **327**, 1470–5.

D'Aversa, E., Bellucci, G., Altieri, F., *et al.* (2011). Spectral characteristics of a spoke on the Saturn Rings. *Mem. S.A.It.* Suppl. **16**, 70.

Deau, E. (2012). Physical properties of the Saturn's rings with the opposition effect. *Geophys. Res. Abstr.*, **14**, 7523–3.

Degiorgio, K., Ferrari, C., Rodriguez, S., and Brahic, A. (2011). Opposition effect of Saturn's rings: Hints of ring physical properties. Paper presented at the EPSC-DPS Joint Meeting 2–7 October 2011, Nantes, France, p. 732.

de Pater, I. and Lissauer, J. J. (2001). *Planetary Sciences*. Cambridge: Cambridge University Press.

(2010). *Planetary Sciences*, 2nd edn. Cambridge: Cambridge University Press.

de Pater, I. Gibbard, S. G., Chiang, E., *et al.* (2005). The dynamic Neptunian ring arcs: Evidence for a gradual disappearance of Liberte and resonant jump of courage. *Icarus*, **174**, 1, 263–272.

de Pater, I., Hammel, H. B., Gibbard, S. G., Showalter, M. R., and van Dam, M. A. (2007). The dark side of the rings of Uranus. *Science*, **317**, 1888–90.

Dermott, S. (1984). Dynamics of narrow rings. In *Planetary Rings*, ed. R. Greenberg and A. Brahic. Tucson, AZ: University of Arizona Press, pp. 589–640.

Descartes, R. (1644). Principia Philosophie. In *Oeuvres de Descartes*, ed. C. Adams and P. Tannery, vol. VIII (Paris, 1905). French edn: vol. IX (Paris, 1904).

Dilley, J. P. (1993). Energy loss in collisions of icy spheres: Loss mechanism and size-mass dependence. *Icarus*, **105**, 225–34.

Dilley, J. P. and Crawford, D. (1996). Mass dependence of energy loss in collisions of icy spheres: An experimental study. *J. Geophys. Res.*, **101**, 9267–70.

Dollfus, A. (1970). Optical reflectance polarimetry of Saturn's globe and rings. *Icarus*, **40**, 171–9.

Dones, H. L. (1991). A recent cometary origin for Saturn's rings? *Icarus*, **92**, 194–203.

Dougherty, M. K., Esposito, L. W., and Krimigis, S. M. (2009). Overview. In *Saturn from Cassini–Huygens*, ed. M. Dougherty, L. W. Esposito, and S. Krimigis. Dordrecht: Springer-Verlag, pp. 3–8.

Dumas, C., Tenile, R. J., Smith, B. A., Schneider, G., and Bectlin, E. E. (1999). Stability of Neptune's ring arcs in question. *Nature*, **400**, 733–5.

Durda, D. D., Greenberg, R., and Jedicke, R. (1998). Collisional models and scaling laws: a new interpretation of the shape of the main-belt asteroid size distribution. *Icarus*, **135**, 431–40.

Durisen, R. H., Cramer, N. L., Mullikin, T. L., and Cuzzi, J. N. (1989). Ballistic transport in planetary ring systems due to particle erosion mechanisms. I. Theory, numerical methods, and illustrative examples. *Icarus*, **80**, 136–66.

Durisen, R. H., Bode, P. W., Cuzzi, J. N., Cederbloom, S. E., and Murphy, B. W. (1992). Ballistic transport in planetary ring systems due to particle erosion mechanisms. II. Theoretical models for Saturn's A- and B-ring inner edges. *Icarus*, **100**, 364–93.

Durisen, R. H., Bode, P. W., Dyck, S. G., *et al.* (1996). Ballistic transport in planetary ring systems due to particle erosion mechanisms. III. Torques and mass loading by meteoroid impacts. *Icarus*, **124**, 220–36.

Elliot, J. L. (1979). Stellar occultation studies of the solar system. *Ann. Rev. Astrophys.*, **17**, 445–75.

Elliot, J. L. and Kerr, R. (1984). *Rings*. Cambridge, MA: MIT Press.

Elliot, J. L., Dunham, E. W., and Mink, D. J. (1977). The rings of Uranus. *Nature*, **267**, 328–30.

Elliott, J. P. and Esposito, L. W. (2011). Regolith depth growth on an icy body orbiting Saturn and evolution of bidirectional reflectance due to surface composition changes. *Icarus*, **212**, 268–74.

Esposito, L. W. (1978) Light scattering from Saturn's rings calculated by a Markov chain formalism. Ph.D. thesis, University of Massachusetts, Amherst, MA.

(1986). Structure and evolution of Saturn's rings. *Icarus*, **67**, 345–57.

(1993). Understanding planetary rings. *Ann. Rev. Earth Planet Sci.*, **21**, 487–523.

(2002). Planetary rings. *Rep. Progr. Phys.*, **65**, 1741–83.

(2010). Composition, structure, dynamics and evolution of Saturn's rings. *Ann. Rev. Earth Planet Sci.*, **38**, 383–410.

Esposito, L. W. and Colwell, J. E. (1989). Creation of the Uranus rings and dust bands. *Nature*, **339**, 637–40.

(1992). Neptune's rings and satellite system: collisional origin and evolution. Paper presented at the American Geophysical Union meeting, Montreal, Canada.

(2003). The effectiveness of cosmic recycling in the history of planetary rings and ring moons. Paper presented at an American Geophysical Union meeting, San Francisco, CA.

Esposito, L. W. and House, L. L. (1978). Radiative transfer calculated from a Markov-chain formalism. *Astrophys. J.*, **219**, 1058–67.

Esposito, L. W. and Lumme, K. (1977). The tilt effect for Saturn's rings. *Icarus*, **31**, 157–67.

Esposito, L. W., Dilley, J. P., and Fountain, J. W. (1980). Photometry and polarimetry of Saturn's rings from Pioneer 11. *J. Geophys. Res.*, **85**, 5948–56.

Esposito, L. W., O'Callaghan, M., and West, R. A. (1983a). The structure of Saturn's rings: implications from the Voyager stellar occultation. *Icarus*, **56**, 439–52.

Esposito, L. W., O'Callaghan, M., Simmons, K. E., *et al.* (1983b). Voyager PPS stellar occultation of Saturn's ring. *J. Geophys. Res.*, **88**, 8643–9.

Esposito, L. W., Cuzzi, J. N., Holberg, J. B., *et al.* (1984). Saturn's rings: structure, dynamics and particle properties. In *Saturn*, ed. T. Gehrels and M. S. Matthews. Tucson, AZ: University of Arizona Press, pp. 463–545.

Esposito, L. W., Harris, C. C., and Simmons, K. E. (1987). Features in Saturn's rings. *Astrophys. J. Suppl.*, **63**, 749–70.

Esposito, L. W., Brahic, A., Burns, J. A., and Marouf, E. A. (1991). Particle properties and processes in Uranus' rings. In *Uranus*, ed. J. T. Bergstralh, E. D. Miner, and M. S. Matthews. Tucson, AZ: University of Arizona Press, pp. 410–68.

Esposito, L. W., Colwell, J. E., and Canup, R. M. (1997). History of Neptune's ring arcs. Paper presented at the 29th Meeting of the Division of Planetary Sciences of the American Astronomical Society, Abstract 17.12. *Bull. Am. Astron. Soc.*, **29**, 1000.

Esposito, L. W., Colwell, J. E., Larsen K., *et al.* (2005). Ultra-violet imaging spectroscopy shows an active Saturn system. *Science*, **307**, 1251–5.

Esposito, L. W., Meinke, B. K., Colwell, J. E., Nicholson, P. D., and Hedman, M. M. (2008). Moonlets and clumps in Saturn's F ring. *Icarus*, **194**(1), 278–89.

Esposito, L. S., Albers, N., Meinke, B. K., *et al.* (2012). A predator–prey model for moon-triggered clumping in Saturn's rings. *Icarus*, **217**(1), 103–14.

Flasar, F. M., Achterberg, R. K., Conrath, B. J., *et al.* (2005). Temperatures, winds, and composition in the Saturnian system. *Science*, **307**, 1247–51.

Foryta, D. W. and Sicardy, B. (1996). The dynamics of the Neptunian Adams ring's arcs. *Icarus*, **123**, 129.

French, R. G. and Nicholson, P. D. (2000). Saturn's rings. II. Particle sizes inferred from stellar occultation data. *Icarus*, **145**, 502–23.

French, R. G., Nicholson, P. D., Porco, C. C., and Marouf, E. A. (1991). Dynamics and structure of the Uranian rings. In *Uranus*, ed. J. T. Bergstralh, E. D. Miner, and M. S. Matthews. Tucson, AZ: University of Arizona Press, pp. 410–68.

French, R. G., McGhee, C. A., Nicholson, P. D., Dones, L., and Lissauer, J. J. (1999). Saturn's wayward shepherds. Paper presented at American Astronomical Society Division of Dynamical Astronomy meeting, Estes Park, CO.

French, R. G., McGhee, C., Dones, L., and Lissauer, J. (2003). Saturn's wayward shepherds: The peregrinations of Prometheus and Pandora. *Icarus*, **162**, 144–71.

French, R. G., Salo, H., McGhee, C. A., and Dones, L. (2007). HST observations of azimuthal asymmetry in Saturn's rings. *Icarus*, **189**, 493–522.

French, R. G., Showalter, M. R., Sfair, R., *et al.* (2012). The brightening of Saturn's F ring. *Icarus*, **219**, 181–93.

Gehrels, T., Baker, L. R., Beshore, E., *et al.* (1980). Imaging photopolarimeter on Pioneer Saturn. *Science*, **207**, 434–9.

Goertz, C. K. and Morfill, G. (1983). A model for the formation of spokes in Saturn's rings. *Icarus*, **53**, 219–29.

Goldreich, P. and Porco, C. C. (1987). The shepherding of the Uranus rings. II. Dynamics. *Astron. J.*, **93**, 730–7.

Goldreich, P. and Rappaport, N. (2003). Origin of chaos in the Prometheus–Pandora system. *Icarus*, **166**, 320–7.

Goldreich, P. and Tremaine, S. D. (1978). The velocity dispersion in Saturn's rings. *Icarus*, **34**, 227–39.

  (1979). Toward a theory for the Uranian rings. *Nature*, **277**, 97–9.

  (1980). Disk–satellite interactions. *Astrophys. J.*, **241**, 425–41.

  (1982). The dynamics of planetary rings. *Ann. Rev. Astron. Astrophys.*, **20**, 249–83.

Goldreich, P., Tremaine, S. D., and Borderies, N. P. (1986). Toward a theory for Neptune's arc rings. *Astron. J.*, **92**, 195–8.

Greenberg, R. (1983). The role of dissipation in the shepherding of ring particles. *Icarus*, **53**, 207–18.

Greenberg, R. and Brahic, A. (eds.) (1984). *Planetary Rings*. Tucson, AZ: University of Arizona Press.

Gresh, D. L. (1990). Voyager radio occultation by the Uranian rings: structure, dynamics and particle size. Ph.D. thesis, Stanford University, Palo Alto, CA.

Gresh, D. L., Marouf, E. A., Tyler, G. L., Rosen, P. A., and Simpson, R. A. (1989). Voyager radio occultation by Uranus's rings. I. Observational results. *Icarus*, **78**, 131–68.

Grün, E., Morfill, G. E., and Mendis, D. A. (1984). Dust–magnetosphere interactions. In *Planetary Rings*, ed. R. Greenberg and A. Brahic. Tucson, AZ: University of Arizona Press, pp. 275–332.

Grün, E., Hamilton, D. P., Baguhl, M., *et al.* (1994). Dust streams from comet Shoemaker-Levy 9? *Geophy. Res. Lett.*, **21**, 1035–8.

Hameen-Anttila, K. A. (1978). An improved and generalized theory for the collisional evolution of Keplerian systems. *Astrophys. Space Sci.*, **58**, 477–519.

(1981). Quasi-equilibrium in collisional systems. *Moon Planets*, **25**, 477–506.

(1982). Saturn's rings and bimodality of Keplerian systems. *Moon Planets*, **26**, 171–96.

Hamilton, D. P. and Burns, J. A. (1994). Origin of Saturn's E ring: self-sustained, naturally. *Science*, **262**, 550–3.

Hamilton, D. P. and Krüger, H. (2008). The sculpting of Jupiter's gossamer rings by its shadow. *Nature*, **453**, 72–5.

Hansen, J. E. and Travis, L. D. (1974) Light scattering in planetary atmospheres. *Space Sci. Rev.*, **16**, 527–610.

Hansen, C. J., Shemansky, D. E., Esposito, L. W., *et al.* (2011). The composition and structure of the Enceladus plume. *Geophys. Res. Lett*, **38**, L11202.

Hapke, B. W., Nelson, R. M., Brown, R. H., *et al.* (2006) Cassini observations of the opposition effect of Saturn's rings. 2 Interpretation: Plaster of Paris as an analog of ring particles. Paper presented at the 37th Annual Lunar and Planetary Science Conference, March 13–17, 2006, League City, TX.

Harris, A. W. (1984). The origin and evolution of planetary rings. In *Planetary Rings*, ed. R. Greenberg and A. Brahic. Tucson, AZ: University of Arizona Press, pp. 641–59.

Hedman, M. M., Burns, J. A., and Tiscareno, M. S., *et al.* (2007a). The source of Saturn's G ring. *Science*, **317**, 653–6.

Hedman, M. M., Nicholson, P. D., and Salo, H., *et al.* (2007b). Self-gravity wake structures in Saturn's A ring revealed by Cassini VIMS. *Astron. J.*, **133**, 2624–9.

Hedman, M. M., Burns, J. A., Showalter, M. R., *et al.* (2007c). Saturn's dynamic D ring. *Icarus*, **188**, 89–107.

Hedman, M. M., Murray, C. D., Cooper, N. J., *et al.* (2009). Three tenuous rings/arcs for three tiny moons. *Icarus*, **199**, 378–86.

Hedman, M. M., Cooper, N. J., Murray, C. D., *et al.* (2010). Aegaeon (Saturn LIII), a G-ring object. *Icarus*, **207**, 433–47.

Hedman, M. M., Burns, J. A., Evans, M. W., Tiscareno, M. S., and Porco, C. C. (2011). Saturn's curiously corrugated C ring. *Science*, **332**, 708–11.

Hedman, M. M., Burns, J. A., Hamilton, D. P., and Showalter, M. R. (2013). Of horseshoes and heliotropes: Dynamics of dust in the Encke Gap. *Icarus*, **223**, 252–76.

Hertzsch, J. M., Spahn, F., and Brilliantov, N. V. (1995). On low-velocity collisions of viscoelastic particles. *J. Phys. II France*, **5**, 1725–38.

Holberg, J. B., Forester, W., and Lissauer, J. J. (1982). Identification of resonance features within the rings of Saturn. *Nature*, **297**, 115–20.

Horanyi, M. (1996). Charged dust dynamics in the solar system. *Ann. Rev. Astron. Astrophys.*, **34**, 383–418.

(1998). Dust plasma interactions at Jupiter. *Astrophys. Space Sci.*, **264**, 257–71.

(2000). Dust streams from Jupiter and Saturn. *Phys. Plasmas*, **7**, 3847.

Horanyi, M. and Burns, J. A. (1991). Charged dust dynamics: Orbital resonance due to planetary shadows. *J. Geophys. Res.*, **96**, A11, 19283–9.

Horanyi, M. and Cravens, T. E. (1996). The structure and dynamics of Jupiter's ring. *Nature*, **381**, 293–5.

Horanyi, M., Hartquist, T. W., Havnes, O., Mendis, D. A., and Morfill, G. E. (2004). Dusty plasma effects in Saturn's magnetosphere. *Rev. Geophys.*, **42**, RG4002.

Horanyi, M., Burns, J. A., Hedman, M. M., Jones, G. H., and Kempf, S. (2009). Diffuse rings. In *Saturn from Cassini–Huygens*, ed. M. Dougherty, L. W. Esposito, and S. Krimigis. Dordrecht: Springer-Verlag, pp. 511–36.

Horanyi, M., Morfill, G. E., and Cravens, T. E. (2010). Spokes in Saturn's B ring: Could lightning be the cause? *IEEE Trans. Plasma Sci.*, **38** (4), 874–9.

Horn, L. J., Yanamandra-Fisher, P. A., Esposito, L. W., and Lane, A. L. (1988). Physical properties of the Uranian delta ring from a possible density wave. *Icarus*, **76**, 485–92.

Housen, K. R. and Holsapple, K. A. (1990). On the fragmentation of asteroids and planetary satellites. *Icarus*, **84**, 226–53.

Housen, K. R., Schmidt, R. M., and Holsapple, K. A. (1991). Laboratory simulations of large scale fragmentation events. *Icarus*, **94**, 180–90.

Hubbard, W. B., Brahic, A., Sicardy, B., *et al.* (1986). Occultation detection of a Neptunian ring-like arc. *Nature*, **319**, 636–40.

Huygens, C. (1656). De Saturni luna observatio nova. In *Oeuvres Complètes de Christiaan Huygens*, vol. 15 (The Hague, 1925), pp. 172–7.

Ip, W.-H. (1988). An evaluation of a catastrophic fragmentation origin of the Saturnian ring system. *Astron. Astrophys.*, **199**, 340–2.

Irvine, W. M. (1975). Multiples scattering in planetary atmospheres. *Icarus*, **25**, 175.

Jerousek, R. G., Colwell, J. E., and Esposito, L. W. (2011). Morphology and variability of the Titan ringlet and Huygens ringlet edges. *Icarus*, **216**(1), 280–91.

Johnson R. E., Luhmann J. G., Tokar R. L., *et al.* (2006). Production, ionization and redistribution of O2 in Saturn's ring atmosphere. *Icarus*, **180**, 393–402.

Jones, G. H., Krupp, N., Krüger H., *et al.* (2006). Formation of Saturn's ring spokes by lightning-induced electron beams. *Geophys. Res. Lett.*, **33** (21), L21202.

Julian, W. H. and Toomre, A. (1966). Non-axisymmetric responses of differentially rotating disks of stars. *Astrophys. J.*, **146**, 810–32.

Keeler, J. E. (1985). A spectroscopic proof of the meteoric constitution of Saturn's rings. *Astron. J.*, **1**, 416–27.

Kemeny, J. and Snell, J. (1960). *Finite Markov Chains*. Princeton, NJ: Van Nostrand.

Kempf, S., Srama, R., Postberg, F., *et al.* (2005). Composition of Saturnian stream particles. *Science*, **307**(5713), 1274–6.

Kolvoord, R. A. and Burns, J. A. (1992). Three-dimensional perturbations of particles in a narrow planetary ring. *Icarus*, **95**, 253–64.

Kolvoord, R. A., Burns, J. A., and Showalter, M. R. (1990). Periodic features in Saturn's F ring. *Nature*, **345**, 675–7.

Krüger, H., Horanyi, M., Krivov, A. V., and Graps, A. L. (2004). Jovian dust: streams, clouds and rings. In *Jupiter: The Planet, Satellites and Magnetosphere*, ed. F. Bagenal, T. Dowling, and W. McKinnon. Cambridge: Cambridge University Press, pp. 219–40.

Lagage, P. O. and Pantin E. (1994). Dust depletion in the inner disk of Beta-Pictoris as a possible indicator of planets. *Nature*, **369**, 628–30.

Landau, L. and Lifshitz, E. M. (1969). *Statistical Physics*, vol. 5. Menlo Park, CA: Addison-Wesley.

Lane, A. L., Hord, C. W., West, R. A., *et al.* (1982). Photopolarimetry from Voyager 2: preliminary results on Saturn, Titan, and the rings. *Science*, **215**, 537–43.

Lewis, M. D. (2001). Dynamics of strongly perturbed planetary rings. Ph.D. thesis, University of Colorado, Boulder, CO.

Lewis, M. D. and Stewart, G. R. (2000). Collisional dynamics of perturbed planetary rings. *Astron. J.*, **120**, 3295–310.

(2005). Expectations for Cassini observations of ring material with nearby moons. *Icarus*, **178**, 124–43.

(2009). Features around embedded moonlets in Saturn's rings: The role of self-gravity and particle size distribution. *Icarus*, **199**, 387–412.

Leyrat, C., Ferrari, C., Charnoz, S., *et al.* (2008). Spinning particles in Saturn's C ring: Pre-Cassini results. *Icarus*, **196**(2), 625–41.

Lin, D. N. C. and Bodenheimer, P. (1981). On the stability of Saturn's rings. *Astrophys. J.*, **248**, L83–6.

Lin, D. N. C. and Papaloizou, J. (1979). Tidal torques on accretion disks in binary systems with extreme mass ratios. *Mon. Not. Roy. Astron. Soc.*, **186**, 799–812.

Lissauer, J. J. (1985). Shepherding model for Neptune's arc ring. *Nature*, **318**, 544–5.

Lissauer, J. J. and Cuzzi, J. N. (1982). Resonances in Saturn's rings. *Astron. J.*, **87**, 1051–8.

Lissauer, J. J. and French, R. G. (2000). HST high-resolution backscatter image of Saturn's G ring. *Icarus*, **146**, 12–18.

Lissauer, J. J. and Sicardy, B. (1990). Models of arcs in Neptune's 63K ring. Paper presented at the 23rd International Council of Scientific Unions Committee on Space Research (COSPAR) meeting, The Hague, Netherlands.

Lissauer, J. J. and Stewart, G. R. (1993). Growth of planets from planetesimals. In *Protostars and Planets III*, ed. E. H. Levy and J. I. Lunine. Tucson, AZ: University of Arizona Press, pp. 106–8.

Lissauer, J. J., Squyres, S. W., and Hartmann, W. K. (1988). Bombardment history of the Saturn system. *J. Geophys. Res.*, **93**, 13776–804.

Longaretti, P.-Y. and Borderies, N. P. (1986). Non-linear study of the Mimas 5 : 3 density wave. *Icarus*, **67**, 211–33.

Lumme, K. A. and Irvine, W. M. (1979). A model for the azimuthal brightness variations in Saturn's rings. *Nature*, **282**, 695–6.

Lumme, K. A., Esposito, L. W., Irvine, W. M., and Baum, W. A. (1977). Azimuthal brightness variations of Saturn's rings. II. Observations at an intermediate tilt angle. *Astrophys. J.*, **216**, L123–6.

Lumme, K., Irvine, W. M., and Esposito, L. W. (1983). Theoretical interpretation of the ground-based photometry of Saturn's B ring. *Icarus*, **53**, 174–84.

Lynden-Bell, D. and Pringle, J. E. (1974). The evolution of viscous discs and the origin of the nebular variables. *Mon. Not. Roy. Astron. Soc.*, **168**, 603–37.

Masset, F. S. and Papaloizou, J. C. B. (2003). Runaway migration and the formation of hot Jupiters. *Astrophys. J.*, **588**(1), 494–508.

Matson, D. L., Spilker, L. J., Lebreton, J.-P., *et al.* (2004). The Cassini/Huygens mission to the Saturnian system. *Space Sci. Rev.*, **104**, 1–58.

McGhee, C. A., French, R. G., Dones, L., and Cuzzi, J. N. (2005). HST observations of spokes in Saturn's B ring. *Icarus*, **173**, 508–21.

McMuldroch, S. M., Pilorz, S. H., and Danielson, G. E. (1999). Galileo NIMS near-IR observations of Jupiter's ring system. *Icarus*, **146**, 1–11.

Meinke, B. K., Esposito, L. W., Albers, N., Sremcevic, M., and Murray, C. (2012). Classification of F ring features observed in Cassini UVIS occultations. *Icarus*, **218**(1), 545–54.

Meinke, B. K., Esposito, L. W, and Stewart, G. R. (2013). Accretion models for Saturn's F ring kittens. (In preparation.)

Meyer-Vernet, N. and Sicardy, B. (1984). On the physics of resonant disk–satellite interaction. *Icarus*, **69**, 157–75.

Mignard, F. (1984). Effects of radiation forces on dust particles. In *Planetary Rings*, ed. R. Greenberg and A. Brahic. Tucson, AZ: University of Arizona Press, pp. 333–66.

Mishchenko, M. I. (1993). On the nature of the polarization opposition effect exhibited by Saturn's rings. *Astrophys. J.*, **411**, 351–61.

Mitchell, C. J., Horanyi, M., Havnes, O., and Porco, C. C. (2006). Saturn's spokes: Lost and found. *Science*, **311**, 1587–9.

Mitchell, C. J., Porco, C. C., Dones, H. L., and Spitale, J. N. (2013). The behavior of spokes in Saturn's B ring. (Submitted.)

Moore, P. (1995). *The Planet Neptune*. Hoboken, NJ: John Wiley.

Morfill, G. E., Fechtig, H., Grün, E., and Goertz, C. K. (1983a). Some consequences of meteoroid impacts on Saturn's rings. *Icarus*, **55**, 439–47.

Morfill, G. E., Grün, E., Johnson, T. V., and Goertz, C. K. (1983b). On the evolution of Saturn's spokes: theory. *Icarus*, **53**, 230–5.

Murray, C. D. and Dermott, S. F. (1999). *Solar System Dynamics*. Cambridge: Cambridge University Press.

Murray, C. D. and Thompson, R. P. (1990). Orbits of shepherd satellites deduced from the structure of the rings of Uranus. *Nature*, **348**, 499–502.

Murray, C. D., Gordon, M. K., and Giuliatti-Winter, S. M. (1997). Unraveling the strands of Saturn's F ring. *Icarus*, **129**, 304–16.

Murray, C. D., Chavez, C., Beurle, K., *et al.* (2005). How Prometheus creates structure in Saturn's F ring. *Nature*, **437**, 1326–9.

Murray, C. D., Beurle, K., Cooper, N. J., *et al.* (2008). The determination of the structure of Saturn's F ring by nearby moonlets. *Nature*, **453**, 739–44.

Namouni, F. and Porco, C. C. (2002). The confinement of Neptune's ring arcs by the moon Galatea. *Nature*, **417**, 45–7.

Nicholson, P. D. and Dones, L. R. (1991). Planetary rings. *Rev. Geophys.*, **29** (suppl.), 313–27.

Nicholson, P. D., Cooke, M. L., Matthews, K., Elias, J. H., and Gilmore, G. (1990). Five stellar occultations by Neptune: Further observations of ring arcs. *Icarus*, **87**, 1–39.

Nicholson, P. D., Mosqueira, I., and Matthews, K. (1995). Stellar occultation observations of Neptune's rings: 1984–1988. *Icarus*, **113**, 295–330.

Nicholson, P. D., Showalter, M. R., Dones, L., *et al.* (1996). Observations of Saturn's ring-plane crossings in August and November 1995. *Science*, **272**, 509–15.

Ockert, M. E., Cuzzi, J. N., Porco, C. C., and Johnson, T. V. (1987). Uranian ring photometry: Results from Voyager 2. *J. Geophys. Res.*, **92**, 14696–979.

Ockert-Bell, M. E., Burns, J. A., Daubar, I. J., *et al.* (1999). The structure of Jupiter's ring system as revealed by the Galileo imaging experiment. *Icarus*, **138**, 188–213.

Ohtsuki, K. (1993). Capture probability of colliding planetesimals: Dynamical constraints on the accretion of planets, satellites, and ring particles. *Icarus*, **106**, 228–46.

Pan, M., Rein, H., Chaing, E., and Evans, S. N. (2012). Stochastic flights of propellers. *Mon. Not. Roy. Astron. Soc.*, **427**(4), 2788–96.

Papaloizou, J. C. B., Nelson, R. P., Kley, W., Masset, F. S., and Artymowicz, P. (2007). *Protostars and Planets V*, ed. B. Reipurth, D. Jewitt, and K. Keil. Tucson, AZ: University of Arizona Press, pp. 655–68.

Peale, S. J. (1986). Orbital resonances, unusual configurations and exotic rotation states among the planetary satellites. In *Satellites*, ed. J. A. Burns and M. S. Matthews. Tucson, AZ: University of Arizona Press, pp. 159–223.

Pollack, J. B. and Cuzzi, J. N. (1980). Scattering by nonspherical particles of size comparable to a wavelength: a new semi-empirical theory and its application to tropospheric aerosols. *J. Atmos. Sci.*, **37**, 868–81.

Porco, C. C. (1983). Voyager observations of Saturn's rings. Ph.D. thesis, California Institute of Technology, Pasadena, CA.

(1990). Narrow rings: observation and theory. *Adv. Space Res.*, **10**, 221–9.

(1991). An explanation for Neptune's ring arcs. *Science*, **253**, 995–1001.

Porco, C. C. and Goldreich, P. (1987). Shepherding to the Uranian rings. I. Kinematics. *Astron. J.*, **93**, 724–9.

Porco, C. C. and Hamilton, D. P. (2007). Planetary rings. In *Encyclopedia of the Solar System*, 2nd edn, ed. L-A. McFadden, P. Weissman, and T. Johnson. San Diego, CA: Academic Press, pp. 503–18.

Porco, C. C. and Nicholson, P. D. (1987). Eccentric features in Saturn's outer C ring. *Icarus*, **72**, 437–67.

Porco, C. C., Nicholson, P. D., Borderies, N., *et al.* (1984a). The eccentric Saturnian ringlets at 1.29R*S* and 1.45R*S*. *Icarus*, **60**, 1–16.

Porco, C. C., Danielson, G. E., Goldreich, P., Holberg, J. B., and Lane, A. L. (1984b). Saturn's non-axisymmetric ring edges at 1.95R*S* and 2.27R*S*. *Icarus*, **60**, 17–28.

Porco, C. C., Nicholson, P. D., Cuzzi, J. N., Lissauer, J. J., and Esposito, L. W. (1995). Neptune's ring system. In *Neptune*, ed. D. P. Cruikshank. Tucson, AZ: University of Arizona Press, pp. 703–806.

Porco, C. C., West, R. A., McEwen, A., *et al.* (2003). Cassini imaging of Jupiter's atmosphere, satellites, and rings. *Science*, **299**, 1541–7.

Porco, C. C., Baker, E., Barbra, J., *et al.* (2005). Cassini imaging science: initial results on Saturn's rings and small satellites. *Science*, **25**(307), 1226–36.

Porco, C. C., Weiss, J. W., Richardson, D. C., *et al.* (2008). Simulations of the dynamical and light-scattering behavior of Saturn's rings and the derivation of ring particle and disk properties. *Astron. J.*, **136**, 2172–200.

Postberg, F., Schmidt J., Hillier, J. K., Kempf S., and Srama, R. (2011). A salt-water reservoir as the source of a compositionally stratified plume on Enceladus. *Nature*, **474**(7353), 620–22.

Poulet, F. and Sicardy, B. (2001). Dynamical evolution of the Prometheus–Pandora system. *Mon. Not. Roy. Astron. Soc.*, **322**, 343–55.

Poulet, F., Sicardy, B., Nicholson, P. D., Karkoschka, E., and Caldwell, J. (2000). Saturn's ring plane crossings of August and November 1995: a model for the new F ring objects. *Icarus*, **144**, 135–48.

Rein, H. and Papaloizou, J. C. B. (2010). Stochastic orbital migration of small bodies in Saturn's rings. *Astron. Astrophys.*, **524**, A22, 13.

Robbins, S. J., Stewart, G. R., Lewis, M. C., Colwell, J. E., and Sremcevic, M. (2010). Estimating the masses of Saturn's A and B rings from high-optical depth *N*-body simulations and stellar occultations. *Icarus* (special online issue), **206**, 431–45.

Roche, E. (1849). Mémoire sur la figure d'une masse fluide soumise à l'attraction d'un point éloigné. *Mém. Acad. Sci. Lett. Montpellier, Sect. Sciences*, **1**, 243–62, 333–48.

Röpke, G. (1987). *Statistische Mechanik für das Nichtgleichgewicht*. Berlin: VEB Deutscher Verlag der Wissenschaften.

Rosen, P. A. (1989). Waves in Saturn's rings probed by radio occultation. Ph.D. thesis, Stanford University, Palo Alto, CA.

Rosen, P. A., Tyler, G. L., and Marouf, E. A. (1991a). Resonance structures in Saturn's rings probed by radio occultation. I. Methods and examples. *Icarus*, **93**, 3–24.

Rosen, P. A., Tyler, G. L., Marouf, E. A., and Lissauer, J. J. (1991b). Resonance structures in Saturn's rings probed by radio occultation. II. Results and interpretation. *Icarus*, **93**, 25–44.

Salmon, J., Charnoz, S., Crida, A., and Brahic, A. (2010). Long-term and large-scale viscous evolution of dense planetary rings. *Icarus*, **209**(2), 771–85.

Salo, H. (1992). Numerical simulations of dense collisional systems. II. Extended distribution of particle sizes. *Icarus*, **96**, 85–106.

(1995). Simulations of dense planetary rings. III. Self-gravitating identical particles. *Icarus*, **117**, 287–312.

(2001). Numerical simulations of collisional dynamics of planetary rings. In *Granular Gases*, ed. T. Pöschel and S. Luding. Berlin: Springer-Verlag, pp. 330–49.

Salo, H. and Hänninen, J. (1998). Neptune's partial rings: action of Galatea on self-gravitating arc particles. *Science*, **282**, 1102–4.

Salo, H. and Karjalainen, R. (2003). Photometric modeling of Saturn's rings. I. Monte Carlo method and the effect of nonzero volume filling factor. *Icarus*, **164**, 428–60.

Salo, H., Schmidt, J., and Spahn, F. (2001). Viscous overstability in Saturn's B ring. I. Direct simulations and measurement of transport coefficients. *Icarus*, **153**, 295–315.

Schaffer, L. E. (1989). The dynamics of dust in planetary magnetospheres. Ph.D. thesis, Cornell University, Ithaca, NY.

Schaffer, L. E. and Burns, J. A. (1987). The dynamics of weakly charged dust: motion through Jupiter's gravitational and magnetic field. *J. Geophys. Res.*, **92**, 2264–80.

Schlichting, H. E. and Chang, P. (2011). Warm Saturns: On the nature of rings of extrasolar planets that reside inside the ice line. *Astrophys. J.*, **734**, 117.

Schmidt, J., Salo, H., Spahn, F., and Petzschmann, O. (2001). Viscous overstability in Saturn's B-ring. II. Hydrodynamic theory and comparison to simulations. *Icarus*, **153**, 316–31.

Schmidt, J., Ohtsuki, K., Rappaport, N., Salo, H., and Spahn, F (2009). Dynamics of Saturn's dense rings. In *Saturn from Cassini–Huygens*, ed. M. Dougherty, L. W. Esposito, and S. Krimigis. Dordrecht: Springer-Verlag, pp. 413–58.

Showalter, M. R. (1989). Anticipated time variations in (our understanding of) Jupiter's ring system. In *Time-Variable Phenomena in the Jovian System*, NASA Special Publication No. 494, ed. M. J. S. Belton, R. A. West, and J. Rahe. Washington, DC: National Aeronautics and Space Administration, pp. 116–25.

(1991). The visual detection of 1981S13 and its role in the Encke Gap. *Nature*, **351**, 709–13.

(1998). Detection of centimeter-sized meteoroid impact events in Saturn's F ring. *Science*, **282**, 1099–102.

(1999). Neptune's misbehaving rings. *Nature*, **400**, 709–10.

(2004). Disentangling Saturn's F ring. I. Clump orbits and lifetimes. *Icarus*, **171**, 356–71.

Showalter, M. R. and Burns, J. A. (1982). A numerical study of Saturn's F-ring. *Icarus*, **52**, 526–44.

Showalter, M. R. and Lissauer, J. J. (2006). The second ring-moon system of Uranus: Discovery and dynamics. *Science*, **311**, 973–7.

Showalter, M. R. and Nicholson, P. D. (1990). Saturn's rings through a microscope: particle size constraints from the Voyager PPS scan. *Icarus*, **87**, 285–306.

Showalter, M. R., Cuzzi, J. N., Marouf, E. A., and Esposito L. W. (1986). Satellite "wakes" and the orbit of the Encke Gap moonlet. *Icarus*, **66**, 297–323.

Showalter, M. R., Burns, J. A., Cuzzi, J. N., and Pollack, J. B. (1987). Jupiter's ring system: new results on structure and particle properties. *Icarus*, **69**, 458–98.

Showalter, M. R., Cuzzi, J. N., and Larson, S. M. (1991). Structure and properties of Saturn's E ring. *Icarus*, **94**, 451–73.

Showalter, M. R., Pollack, J. B., Ockevt, M. E., Doyle, L. R., and Dalton, J. B. (1992). A photometric study of Saturn's F ring. *Icarus*, **100**, 394–411.

Showalter, M. R., Hedman, M. M., and Burns, J. A. (2011). The impact of comet Shoemaker–Levy 9 sends ripples through the rings of Jupiter. *Science*, **332**(6030), 711–13.

Shu, F. H. (1984). Waves in planetary rings. In *Planetary Rings*, ed. R. Greenberg and A. Brahic. Tucson, AZ: University of Arizona Press, pp. 513–61.

Shu, F. H. and Stewart, G. R. (1985). The collisional dynamics of particulate disks. *Icarus*, **62**, 360–83.

Shu, F. H., Yuan, C., and Lissauer, J. J. (1985a). Nonlinear spiral density waves: an inviscid theory. *Astrophys. J.*, **291**, 356–76.

Shu, F. H., Dones, L. R., Lissauer, J. J., Yuan, C., and Cuzzi, J. N. (1985b). Nonlinear spiral density waves: viscous damping. *Astrophys. J.*, **299**, 542–73.

Sicardy, B. and Lissauer, J. J. (1992). Dynamical models of the arcs in Neptune's 63K ring (1989N1R). *Adv. Space Res.*, **12** (11), 97–111.

Sicardy, B., Roddier, F., Roddier, C., *et al.* (1999). Images of Neptune's ring arcs obtained by a ground-based telescope. *Nature*, **400**, 731–3.

Smith, B. A., Soderblom, L., Batson, R., and Suomi, V. E. (1982). A new look at the Saturn system: the Voyager 2 images. *Science*, **215**, 504–37.

Smith, B. A., Soderblom, L. A., Beebe, R., *et al.* (1986). Voyager 2 in the Uranian system: imaging science results. *Science*, **233**, 43–64.

Smith, B. A., Soderblom, L. A., Banfield, D., *et al.* (1989). Voyager 2 at Neptune: imaging science results. *Science*, **246**, 1422–49.

Smith, W. H., McCord, T. B., and Macy, W. (1981). High-spectral-resolution imagery of Saturn. *Icarus*, **46**, 256–62.

Soter, S. (1971). *The Dust Belts of Mars*, Center for Radiophysics and Space Research. Report No. 462. Ithaca, NY: Cornell University.

Spahn, F. M. and Sremcevic, M. (2000). Density patterns induced by small moonlets in Saturn's rings. *Astron. Astrophys.*, **358**, 368–72.

Spahn, F., Petzschmann, O., Schmidt, J., Sremcevic, M., and Hertzsch, J.-M. (2001). Granular viscosity, planetary rings and inelastic particle collisions. In *Granular Gasses*, ed. T. Pschel and S. Luding. Berlin: Springer-Verlag, pp. 363–85.

Spahn, F., Albers, N., Sremcevic, M., and Thornton, C. (2004). Kinetic description of coagulation and fragmentation in mesoscopic particle ensembles. *Europhys. Lett.*, **67**, 545–51.

Spilker, L. J., Pilorza, S., Lanea, A. L., *et al.* (2004). Saturn A ring surface mass densities from spiral density wave dispersion behavior. *Icarus*, **171**, 372–90.

Spitale, J. and Porco, C. (2007). Association of the jets of Enceladus with the warmest regions on its south-polar fractures. *Nature*, **449**, 695–7.

Sremcevic, M., Schmidt, J., Salo, H., *et al.* (2007). A belt of moonlets in Saturn's A ring. *Nature*, **499**, 1019–21.

Sremcevic, M., Colwell, J. E. and Esposito, L. W. E. (2013). UVIS tracking occultations. (In preparation.)

Stevenson, D. J., Harris, A. W., and Lunine, J. I. (1986). Origins of satellites. In *Satellites*, ed. J. A. Burns and M. S. Matthews. Tucson, AZ: University of Arizona Press, pp. 39–88.

Stewart, G. R., Lin, D. N. C., and Bodenheimer, P. (1984). Collision-induced transport processes in planetary rings. In *Planetary Rings*, ed. R. Greenberg and A. Brahic. Tucson, AZ: University of Arizona Press, pp. 447–512.

Stewart, G. R., Robbins, S. J., and Colwell, J. E. (2007). Evidence for a primordial origin of Saturn's rings. Paper presented at the 39th Meeting of the Division for Planetary Sciences of the American Astronomical Society, Abstract. No. 7.06.

Tagger, M., Henriksen, R. N., and Pellat, R. (1991). On the nature of spokes in Saturn's rings. *Icarus*, **91**, 297–314.

Tamayo, D., Burns, J. A., Hamilton, D. P., and Hedman, M. M. (2011). Finding the trigger to Iapetus' odd global albedo pattern: Dynamics of dust from Saturn's irregular satellites. *Icarus*, **215**(1), 260–78.

Thiessenhusen, K.-U., Krüger, H., Spahn, F., and Grün, E. (2000). Dust grains around Jupiter: The observations of the Galileo dust detector. *Icarus*, **144**, 89–98.

Thomas, G. E. and Stamnes, K. (1999). *Radiative Transfer in the Atmosphere and Ocean*. Cambridge: Cambridge University Press.

Thomas, P. C. (1989). The shapes of small satellites. *Icarus*, **77**, 248–74.

Throop, H. B. and Esposito, L. W. (1998). G ring particle sizes derived from ring plane crossing observations. *Icarus*, **131**, 152–66.

Throop, H. B., Porco, C. C., West, R. A., *et al.* (2004). The jovian rings: New results derived from Cassini, Galileo, Voyager, and Earth-based observations. *Icarus*, **172**, 59–77.

Tiscareno, M. S. (2013a). A modified "Type I migration" model for propeller moons in Saturn's rings. *Planetary and Space Science*, **77**, 136–42.

  (2013b). Planetary rings. In *Planets, Stars and Stellar Systems*, ed. P. Kalas and L. French. Berlin: Springer-Verlag, pp. 61–3.

Tiscareno, M. S., Burns J. A., Hedman M. M., *et al.* (2006a). 100-metre-diameter moonlets in Saturn's A ring from observations of "propeller" structures. *Nature.*, **440**, 648–50.

Tiscareno, M. S., Nicholson, P. D., Burns, J. A., Hedman, M. M., and Porco, C. C. (2006b). Unravelling temporal variability in Saturn's spiral density waves: Results and predictions. *Astrophys. J.*, **651**, L65–L68.

Tiscareno, M. S., Burns, J. A., Nicholson, P. D., Hedman, M. M., and Porco, C. C. (2007). Cassini imaging of Saturn's rings: II. A wavelet technique for analysis of density waves and other radial structure in the rings. *Icarus*, **189**, 14–34.

Tiscareno, M. S., Burns, J. A., Hedman, M. M., and Porco, C. C. (2008). The population of propellers in Saturn's A ring. *Astron. J.*, **135**(3), 1083–91.

Tiscareno, M. S., Burns, J. A., Sremcevic. M., *et al.* (2010). Physical characteristics and non-keplerian orbital motion of "propeller" moons embedded in Saturn's rings. *Astrophys. J. Lett.*, **718**, L92–L96.

Toomre, A. (1964). On the gravitational stability of a disk of stars. *Astrophys. J.*, **139**, 1217–38.

Trulsen, J. (1971). Towards a theory of jet streams. *Astrophys. Space Sci.*, **12**, 329–48.

  (1972a). Numerical simulation of jet streams. I. The three-dimensional case. *Astrophys. Space Sci.*, **17**, 241–62.

  (1972b). Numerical simulation of jet streams. II. The two-dimensional case. *Astrophys. Space Sci.*, **18**, 3–20.

Tyler, G. L., Sweetnam, D. N., Anderson, J. D., *et al.* (1986). Voyager 2 radio science observations of the Uranian system: Atmospheres, rings, satellites. *Science*, **233**, 79–84.

Van Allen, J. A. (1982). Findings on rings and inner satellites of Saturn by Pioneer 11. *Icarus*, **51**, 509–27.

  (1983). Absorption of energetic protons by Saturn's G ring. *J. Geophys. Res.*, **88**, 6911–18.

Van Allen, J. A., Thomsen, M. F., Randall, B. A., Rairden, R. L., and Grosskreutz, C. L. (1980). Saturn's magnetosphere, rings, and inner satellites. *Science*, **207**, 415–21.

van de Hulst, H. C. (1957). *Light Scattering by Small Particles*. New York: John Wiley. Revd. edn 1981. New York: Dover Publications.

van Helden, A. (1984). Rings in astronomy and cosmology 1600–1900. In *Planetary Rings*, ed. R. Greenberg and A. Brahic. Tucson, AZ: University of Arizona Press, pp. 12–24.

Wagener, R. and Caldwell, J. (1988). On the abundance of micron-sized particles in Saturn's A and B rings. *ESA SP-281*, **1**, 85.

Ward, W. R. (1981). On the radial structure of Saturn's ring. *Geophys. Res. Lett.*, **8**, 641–3.

Ward, W. R. and Hahn, J. M. (2000). Disk–planet interactions and the formation of planetary systems. In *Protostars and Planets IV*, ed. V. Mannings, A. Boss, and S. Russell. Tucson, AZ: University of Arizona Press, pp. 1135–55.

Weidenschilling, S. J., Chapman, C. R., Davis, D. R., and Greenberg, R. (1984). Ring particles: collisional interactions and physical nature. In *Planetary Rings*, ed. R. Greenberg and A. Brahic. Tucson, AZ: University of Arizona Press, pp. 367–415.

Wisdom, J. and Tremaine, S. D. (1988). Local simulations of planetary rings. *Astron. J.*, **95**, 925–40.

Zebker, H. A., Tyler, G. L., and Marouf, E. A. (1983). On obtaining the forward phase functions of Saturn ring features from radio occultation observations. *Icarus*, **56**, 209–28.

Index

A ring. *See under* Saturn, rings
accretion. *See under* particles
Adams ring. *See under* Neptune, rings
age of rings, 118, *133*
albedo, *155, 157, 220*
alpha (α). *See under* Uranus, rings
Amalthea. *See under* Jupiter, moons
American Astronomical Society's Division for
    Planetary Sciences, *156*
Arago. *See under* Neptune, rings
Araki, Suguru, *11*
arcs, 1, 2, 3, 23, 26, 28, 69, 94, 97, 111, 120, 124, 125,
    156, 157, 159, 160, 161, 162, 163, 164–70,
    171, 195, 196, 198, 199, 202, 205, 229, 230,
    232, 235. *See also* Neptune, arcs
astrophysical approach, 16
Atlas. *See under* Saturn, moons
azimuthal structure, *4, 68, 95, 110, 120, 124*

B ring. *See under* Saturn, rings
Babinet's paradox, 181
ballistic transport, 70, 220
Barbara, John, 126, 132, *146, 149,* 150, 173
bending waves, 67. *See also* waves
beta (β). *See under* Uranus, rings
Boltzmann equation, 51, 54, 220, 221, 223
bombardment by meteorites, 45
Borderies (Rappaport), Nicole, 15, 143
Bosh, Amanda, 97
Brahic, Andre, 157
braids, 95, 98, 136, 220
Bridges, Frank, 36
Brooks, Shawn, 173, 176, 177
Brophy, Thomas, 113
Brownian motion, 112
Burns, Joseph, 4, 120, 149, 150, 195
burst events, 149

C (crepe) ring. *See under* Saturn, rings
Callisto. *See under* Jupiter, moons

Canada–France Hawaii Telescope, 167
Canup, Robin, 129, 177
Cassini Division. *See under* Saturn, rings
Cassini spacecraft, 1–5, 6, 7, 10, 16, 17–18, 20, 30,
    32, 45–7, 48, 53, 60, 62, 63, 66, 70–90, 93,
    107, 110, 118, 126, 130, 133, 134, 137–8,
    144–7, 149, 150, 153, 154, 166, 174, 175, 177,
    179, 180, 182, 183, 185, 190, 194–202, 204,
    205–6, 207, 208, 209–19
    Composite IR Spectrometer (CIRS), 47
    Cosmic Dust Analyzer (CDA), 180, 200
    Ultraviolet Imaging Spectrograph (UVIS), 74, 75,
        84, 146
    Visual and Infrared Mapping Spectrometer (VIMS),
        45, 46, 144, 146, 185
Cassini, Jean Dominique, 10, 39
characteristics. *See* general properties of rings
Chiang, Eugene, 92
coefficient of restitution, 35, 36, 38, 51, 102, 104, 108
collisional cascade, 114, 122, 123, 127, 128, 130, 131,
    132, 133, 134, 221
Colwell, Joshua, 120, 121, 122, 123, 124, 126, 128,
    130, 133, 157, 161
comet Shoemaker-Levy 9, 128, 193
comparison of ring systems, 20–30
composition. *See* general properties of rings
confinement of rings. *See* moons
Cook, Alan, 50
Copernican model, 6
Cordelia. *See* Uranus
corotation resonance. *See under* resonance
cosmic recycling, x, 125, 133, 134, 150, 171, 221, 230
Courage. *See under* Neptune, moons
Cravens, Tom, 178, 191
crepe ring, 29. *See under* Saturn, rings
Cuzzi, Jeffrey, xiv, 3, 45, 46, 55, 90, 96, 120, 126,
    140, 147, 148, 149, 150, 175, 195

D ring. *See under* Saturn, rings
de Pater, Imke, 151, 167

delta (δ). *See under* Uranus, rings

density waves, 3, 18, 22, 29, 30, 55, 58, 63, 66, 67, 73, 77, 80, 85, 86, 126, 128, 130, 133, 141, 204, 206, 211, 238. *See also* waves

Descartes, Rene, 8

Descartes, Rene, 8

"dirt", 46, 211

discovery of rings, 6, 11. *See also individual planet names' rings other than Saturn's*
  Saturn's rings, 6, 136

dissipation importance in resonance, 57

Dollfus, Andouin, 70, 71, 229

Dones, Luke, 128

drag, 2, 16, 27, 58, 92, 93, 116, 119, 120, 127, 131, 188, 190, 191–2, 193, 198, 202

drunkard's walk. *See* random walk

Dumas, Christophe, 163

Durisen, Richard, 90

dust, 14, 20, 22, 40, 44, 76, 92–3, 98, 114, 120, 127, 129, 145, 149, 152, 153, 155, 157, 161–2, 170, 173, 187–202
  clouds, 192–3
  forces on, 187–91
  lifetime of, 191–2, 193

dusty rings. *See* dust

dynamic ephemeral bodies (DEBs), 41, 42, 122

dynamics. *See under* general properties of rings

E ring. *See under* Saturn, rings

Egalité. *See under* Neptune, arcs

elasticity, 51, 105, 108

electromagnetic effects, 87

Elliot, Jim, 12, 16, 156

embedded moons. *See* moons, embedded

Enceladus. *See under* Saturn, moons

Encke Gap. *See under* Saturn, rings

Epimetheus. *See under* Saturn, moons

epsilon (ε) ring. *See under* Uranus, rings

equipartition of energy, 41

Estrada, Paul, 45

eta (η). *See under* Uranus, rings

ethereal rings, 29, 49, 154, 178

Europa. *See under* Jupiter, moons

evolution, 15, 17, 18, 48, 50, 51, 52, 68, 70, 92, 101, 108, 111, 113, 114, 116–34, 141, 150, 162, 170, 172, 173, 178, 179, 184, 187, 192, 204, 206, 209

F ring. *See under* Saturn, rings

flattening, 54

Fomalhaut b, 32

Foryta, Dietmar, 160

Franklin, Fred, 50

Fraternité. *See under* Neptune, arcs

French, Richard, 141

G ring. *See under* Saturn, rings

Galatea. *See under* Neptune, moons

Galilean satellites. *See under* Jupiter

Galilei, Galileo, 6–10

Galileo spacecraft, 16, 24, 26, 92, 93, 94, 95, 126, 154, 172, 173, 174, 175, 176, 177, 192–3, 202, 205

Galle. *See under* Neptune, rings

gamma (γ). *See under* Uranus, rings

Ganymede. *See under* Jupiter, moons

gas drag. *See* drag

Gehrels, Tom, 14, 135–7

general properties of rings
  characteristics, 21, 22, 35, 48, 206
  composition, 1, 3, 25, 34, 180, 184–5
  dynamics, 3, 15, 18, 53, 54, 55, 58, 72, 101, 110, 111, 112, 114, 134, 182, 183, 204, 206
  structure, 2, 3, 4, 5, 15, 17–18, 20–8, 60, 70–100, 93–4, 98, 109–10, 137–8, 146, 150, 204

geophysical approach, 16

Goldreich, Peter, 12, 15, 48, 55, 57, 58, 66, 69, 91, 92, 94, 95, 118, 119, 120, 125, 137, 138, 143, 152, 156, 157, 159, 162, 163

gossamer rings, 24, 25, 26, 93, 94, 172–3, 192, 205

granola bar, 77, 78

Greenberg, Richard, 10

Grün, Eberhard, 88

halo. *See under* Jupiter, rings

Hameen-Anttila, K. A., 55

Hänninen, Jyrki, 161

Hansen, James, 180

Harris, Alan, 116

Herschel, William, 11

Hill sphere, 149

Horanyi, Mihaly, 90, 178

Hubble Space Telescope, 2, 7, 9, 31, 32, 45, 77, 90, 96, 99, 139, 140, 142, 151, 152, 162, 163, 164, 166, 167

Huygens, Christiaan, 9, 10

hydrodynamic approach, 48, 52–3

ice. *See* water ice

inner Lindblad resonance (ILR). *See* resonance

inner vertical resonance (IVR). *See* resonance

instability. *See* stability

International Astronomical Union, 136

Io. *See under* Jupiter, moons

Ip, Wing, 116

Irvine, William, 83, 180

Janus. *See under* Saturn, rings

Jeffreys, Harold, 48

jets, 17, 137, 145, 146, 150, 173, 200

Jupiter, 1, 3, 6, 9, 12–14, 16, 20–5, 26, 32, 33, 44, 58, 92, 93, 94, 95, 97, 118, 128, 135, 154, 155, 156, 157, 172–8, 184, 189–94, 196, 202, 205
  Galilean satellites, 6, 9, 193, 202
  moons, i, 21, 25, 172–8
    Adrastea, 20, 22, 24, 25, 92, 172, 173, 192

Amalthea, 20, 22, 24, 25, 26, 44, 93, 172, 173, 174, 184
Callisto, 6, 193
Europa, 6, 193
Ganymede, 6, 193
Io, 6
Metis, 20, 24, 25, 172, 173, 190
Thebe, 20, 22, 24, 25, 26, 44, 93, 172, 173, 178, 184, 192
rings, 1, 3, 13–14, 16, 20–5, 26, 44, 92–3, 94, 95, 97, 135, 154, 155, 156, 172–8, 184, 190, 191, 192, 193–4, 196, 202, 205
age, 118, 178
composition, 25, 44
discovery, 12–14, 172
dusty (outer, tenuous), 194
halo, 22, 24, 25, 26, 92, 94, 172, 191
Main Ring, 22, 24, 25, 26, 92, 93, 95, 97, 172–8, 191, 205
Metis Notch, 24

Kant, Immanuel, 10
Karjalainen, Raine, 182, 183
Keck telescope, 151, 153, 167–70
Keeler Gap. *See under* Saturn, rings
Keeler, James, 11
Kepler, Johannes, 48
kinetic theory, 48, 51–2
kittens, 80, 145, 150, 173
Kuiper Airborne Observatory, 12, 13, 151

Lagrange points, 69
Laplace, Pierre Simon, 11
Lassell. *See* Neptune
Lassell, William, 28
LeVerrier. *See under* Neptune, rings
Lewis, Mark, 38, 52, 55, 66
Liberté. *See under* Neptune, arcs
libration, 62, 69, 141, 156, 157, 160
Lindblad resonance. *See* resonance
Liouville equation, 114
Lissauer, Jack, 52, 69, 116, 124, 141, 151, 156
longitudinal confinement. *See* azimuthal structure
Lorentz resonances. *See* resonance
Lowell Observatory, 11
Lumme, Kari, 83, 182
Lynden-Bell, Donald, 50

Main Ring. *See under* Jupiter, rings
Markov process/Markov chain, 112–14, 122, 130–1
Maxwell, James Clerk (J.C.), 11, 39, 48
meteor/meteoroid/micrometeoroid impacts, 3, 16, 18, 50, 70, 90, 92, 96, 110, 116, 117, 118, 120, 124, 125, 126, 127, 130, 133, 146, 149, 172, 177, 184, 192, 193, 202, 205, 206
Metis. *See under* Jupiter, moons
Metis Notch. *See under* Jupiter, rings

microstructure, 83, 84, 109
Mie, Gustav/Mie scattering, 174, 180
Millis, Robert, 12
Mimas. *See under* Saturn, moons
Miranda. *See under* Uranus, moons
Mishchenko, Michael, 180
Mitchell, Colin, 90
monolayer, 41, 42, 182
Monte Carlo simulations, 113–14, 121, 122, 183
moonlets, 18, 40, 42, 43, 79–82, 96, 110, 120, 123, 137, 146–7, 150, 154, 156, 160–2, 165, 170, 195, 214
moons. *See also individual moon names and related planets*
confinement of rings, 55–8, 68–9, 95, 141, 160, 165, 166
embedded, 2, 15, 17, 29, 33, 40, 42, 45, 57, 73, 156, 161, 171, 172, 195, 206
source of ring material, 25, 92, 96, 124, 127, 128, 172, 173, 192, 204, 205
Murray, Carl, 61, 64, 124, 137, 145

Namouni, Fathi, 166–70
N-body simulations, 55, 101–2, 103, 104–7, 110, 111, 113, 204
Neptune
arcs, 125
Courage, 28, 157, 164, 169
Egalité, 28, 157, 159, 163, 164, 165, 167
Fraternité, 28, 157, 159, 164, 167, 169, 170
Liberté, 28, 157, 159, 164, 165, 169, 170
discovery, 28
moons, 121–5
Despina, 164, 167
Galatea, 28, 95, 120, 124, 157, 160, 162, 164, 166–7, 169, 170
Larissa, 163, 164, 167
Proteus, 163, 167
rings, 28, 29, 69, 93, 94, 97, 113, 116, 117, 120, 155–70, 174–8
Adams ring, 28, 29, 69, 94, 97, 120, 157, 160, 164, 166, 167, 168
age, 120, 157
Arago, 28
discovery, 162–6
Galle, 28
Laselle, 28
LeVerrier, 28, 97
nomenclature, 25
Triton, 28
New Horizons, 173
Nicholson, Philip, 159–60, 165, 167, 170

occultations, 11, 13, 28, 69, 74–75, 80, 86, 97, 119, 128, 145, 146, 150, 155, 157, 181
Ophelia. *See under* Uranus, moons
orbital velocity, 11, 76, 90, 102
overstability. *See* stability
Owen, Tobias, 13

Pan. *See under* Saturn, moons

Pandora. *See under* Saturn, moons

partial rings. *See* Neptune, arcs *or* Saturn, rings, Encke Gap ringlets

particles, 42. *See also* general properties of rings, composition, dust *and* Saturn, rings

  accretion/reaccretion, 36, 41–4, 50, 51, 79, 82, 111, 113, 114, 120, 121, 122, 128, 130, 131, 134, 147, 149–50, 161, 170, 183, 205, 206

  collisions, 16, 26, 35–8, 40, 41, 49–51, 54, 55, 57, 58, 63, 69, 70, 77, 81, 90, 92, 95, 96, 101–2, 104–8, 110, 113, 114, 118, 120, 122, 125, 126, 141, 149, 150, 155, 156, 157, 160–2, 166, 170, 173, 176, 177, 184, 191, 202, 204, 206

  composition, 2, 25, 34, 44–7, 134, 184, 185, 200, 206

  distribution. *See* particle, size

  dynamics, 54–5

  energy equipartition, 41, 104

  regolith, 36, 80, 81, 114, 126, 130, 149, 173, 183, 184, 204

  size, 1, 2, 3, 4, 14, 17, 24, 27, 34, 35, 36, 39–41, 42, 51, 53, 55, 73, 78, 79–82, 87, 89, 92, 95, 99, 104, 105, 106, 110, 116, 117, 118, 119, 122, 123, 125, 126, 127, 128, 130–1, 134, 138, 145, 146, 149–50, 152, 160, 161, 174–8, 180–2, 184, 187–90, 193, 194, 200, 205, 206

  velocity, 11, 18, 35–6, 48, 49, 50, 51, 54, 58, 65, 73, 88, 102, 103, 104, 106, 107, 108, 109, 117, 122, 123, 161, 190, 192

pendulum model, 61–3

periodic boundary conditions, 102–3

phase function, 174, 181, 185

Phoebe. *See under* Saturn, moons

photometry, 3, 18, 180–4

Pioneer space missions, 6, 11, 12–15, 18, 29, 83, 95, 128, 135–6, 147, 172, 195

plasma drag. *See* drag

Pollack, James, 175

pollution, 46, 114, 117, 126, 129

Porco, Carolyn, 58, 77, 119, 120, 124, 160, 161, 162, 166–7, 170, 200

Poulet, Francois, 141

Poynting–Robertson effect, 131, 190, 192, 193

Pringle, James, 50

Prometheus. *See under* Saturn, moons

propeller, 17–18, 73, 79–82, 110

properties. *See* general properties of rings

radiation belts, 13, 14

random walk, 50, 54, 112

Rappaport. *See* Borderies

recycling of ring material, 16, 17, 30, 79, 125–34, 150, 154, 170, 171, 172, 177, 203, 205

regolith. *See under* particles

resonance, 15, 18, 33, 45, 55–8, 60–3, 64, 66–9, 70–3, 76, 85, 92, 94, 95, 114, 119, 120, 125, 126, 127, 130, 133, 140, 141, 156, 157, 160, 161–2, 165, 166, 167, 170, 172, 173, 178, 187, 191, 194, 195, 202, 204

  corotation resonance, 69, 157, 170, 194

  Lindblad resonance, 61, 62, 64, 69, 140, 156, 160, 173

  Lorentz resonance, 173, 191

  vertical resonance, 61, 62

ring arcs. *See* arcs

ring photometry. *See* photometry

ring plane crossing, 7, 99, 137, 141, 194

ring–moons, 20, 25, 30, 116, 124, 126, 130, 133, 151, 172–4, 190, 205

Roche limit, 1, 20, 21, 33, 42–4, 114, 118, 126, 129, 147, 149, 170

Roche zone, 44, 113, 116, 122, 126, 128, 130, 206

Roche, Edouard, 42

Röpke, Gerd, 114

Rosalind. *See under* Uranus, moons

rubble piles, 41–2, 44, 53, 149, 174, 185, 204

Salo, Heikki, 104, 111, 161–2, 165, 170, 182

satellite disruptions, 121–3

satellites. *See* moons *or* particles

Saturn

  (Saturn) electrostatic discharge (SED), 87

  (Saturn) kilometric radiation (SKR), 87

  moons

    Atlas, 64, 126, 138

    Daphnis, 2

    Dione, 9, 200

    Enceladus, 22, 29, 93, 98, 190, 194, 200

    Epimetheus, 30, 31, 32, 33, 64, 87, 139, 162, 213

    Janus, 32, 64, 86, 87, 139, 162, 213

    Mimas, 29, 55, 62, 63, 73, 86, 118, 128, 133, 141, 194, 195, 200

    new moons, 2, 15, 157

    Pan, 2, 66, 73, 76, 78, 80, 83, 84, 94, 95, 123, 167, 198, 199, 215

    Pandora, 15, 63, 64, 96, 126, 133, 137, 139, 140, 141, 142, 143, 146, 148, 150, 217

    Phoebe, 20, 34, 45, 205

    Prometheus, 15, 18, 63, 64, 96, 126, 133, 137–41, 142, 143, 145, 146, 148, 150, 214, 218

    Rhea, 200

    Tethys, 141, 200

    Titan, 9, 10, 118, 129, 200, 208

  rings, 1–2, 4, 5, 6–11, 14–15, 16, 17–18, 20, 21, 29–30, 31, 32, 34, 39, 41, 42, 44, 45–7, 49, 60, 62, 63, 64, 66, 67, 68, 70–90, 93, 95, 98, 99, 116, 117, 123, 126, 128–30, 133–4, 135–50, 180, 182, 183, 184, 185, 189, 194–202, 205, 207, 208, 209, 210, 211, 212, 213, 214, 215, 216, 217, 218, 219

    A ring, 29–30, 31, 45, 46, 47, 62, 63, 64, 66, 70, 73–7, 80, 82, 83, 84, 85, 86, 94, 96, 106, 107, 110, 117, 118, 126, 139, 141, 167, 198, 209, 211

    age, 118

B ring, 18, 30, 39, 45, 47, 49, 78, 80, 82, 84, 85, 87, 88, 109, 110, 128, 133, 134, 182, 200, 201, 204, 211, 216, 217

C (crepe) ring, 11, 29, 30, 45, 47, 73, 79, 82, 209, 211

Cassini Division, 10, 22, 30, 45, 46, 47, 73, 75, 76, 77, 79, 82, 99, 153, 209

crepe ring. *See* Saturn, rings, C ring

D ring, 73, 194, 196, 197, 202

discovery, 1, 6, 10, 11, 14, 20, 29, 34, 95, 136, 147

E ring, 32, 72, 93, 98, 189, 190, 194, 200

Encke Gap *and* Encke Gap ringlets, 30, 66, 73, 76, 78, 80, 82, 84, 94, 95, 96, 123, 154, 167, 198, 215

F ring, 4, 7, 15, 16, 17, 18, 29, 30, 31, 66, 73, 80, 94–7, 98, 99, 114, 117, 120, 126, 134, 135–50, 152, 165, 170, 173, 194, 204, 205, 214, 218: strand F-α, 137; strand F-β, 137; strand F-δ, 137; strand F-γ, 137

G ring, 29, 31, 32, 93, 97, 98, 133, 180, 194, 196, 198

Keeler Gap, 30, 66, 78, 80, 123, 219

mass, 15, 17, 18, 29, 41, 45, 50, 63, 67, 70, 73, 86–7, 99, 100, 111, 118, 128, 130, 133, 134, 205, 207

nomenclature, 14, 29–30, 136

orbital period, 85, 108, 117, 191, 204

orbital speed. *See* Saturn, rings, orbital period

structure, 2, 3, 4, 5, 15, 17–18, 39, 60, 70, 71, 72–100, 109–10, 137–8, 146, 150, 197, 204

width, 7, 22, 85

seasonal changes, 6, 18, 90, 182, 183, 184, 200

scalloped edge, 15, 66, 204

scattering of light, 180–4, 187, 189

Schmidt, Jüergen, 200

Schmidt, Jürgen, 52

self gravity wakes, 4, 17, 18, 76–82, 110, 215

shepherd/shepherding, 12, 15, 16, 26, 28, 30, 55, 56, 57, 58, 63–6, 96, 116, 117–18, 119–20, 124, 125, 127, 138–44, 146, 148, 152, 204, 205

Shoemaker-Levy 9 comet. *See* comet Shoemaker-Levy 9

Showalter, Mark, 66, 99, 137, 138, 146, 149, 151, 155, 162, 174, 177

Shu, Frank, 15, 48, 55, 58

Sicardy, Bruno, 124, 141, 160, 163

Smith, Bradford, 88, 121, 136

Soter, Steven, 122

Southwest Research Institute, 128, 129, 166

Space Sciences Institute, 2

Spahn, Frank, 37

spiral bending waves. *See* bending waves

spiral density wave. *See* density waves

Spitzer Space Telescope, 20, 32, 34

spokes, 2, 4, 18, 29, 87–90, 93, 111, 194, 200–2, 204, 217

spreading, 50–1, 54, 56, 57, 60, 66, 70, 102, 116, 117, 118, 119, 124, 130, 133, 152

sputtering, 118, 125, 130, 131, 192, 193, 201

stability, 36, 48, 106, 107–10, 165, 166

Stamnes, Knut, 180

stellar occultation. *See* occultations

Stevenson, David, 122

Stewart, Glen, 38, 48, 52, 55, 66, 111, 150

stochastic models and processes, 18, 101, 111–14, 121, 122, 123, 127, 131, 150, 170, 171, 177, 186, 191, 203, 204, 206

strands, 18, 136–7, 145–50

stress tensor, 52

structure. *See* general properties of rings

SuperWASP telescope, 33

temperature, 4, 35, 36, 41, 44–7, 48, 52, 107–8, 110, 182, 185

Terrile, Rich, 11

Tethys. *See under* Saturn, moons

Thebe. *See under* Jupiter, moons

thermal radiation, 182

thickness. *See* optical depth

Thomas, Gary, 180

Thompson, Robert, 124

Throop, Henry, 98, 133, 174, 177

tilt effect, 182, 184

Tiscareno, Matthew, 80, 86, 87

Titan. *See under* Saturn, moons

Tomasko, Martin, 135

Toomre, Alar, 79, 106, 108

torque, 16, 56–8, 67, 96, 119–20, 140

transition matrix, 113, 130, 132

Travis, Larry, 180

Tremaine, Scott, 12, 15, 55, 57, 66, 84, 91, 138, 139, 152

Triton. *See under* Neptune, moons

ultraviolet (UV), 4, 15, 45, 209, 211, 215

Uranus, 11–12, 16, 22, 25–7, 28, 56, 58, 120, 151–4

moons, 121, 151–4

Cordelia, 23, 28, 56, 58, 119, 121, 124, 153, 162

Miranda, 58

Ophelia, 28, 56, 58, 119, 120, 121, 153

Rosalind, 23, 121, 124

rings, 1, 2, 3, 11–12, 13, 25–7, 28, 44, 54, 56, 58, 66, 91, 92, 93, 94, 116–20, 121, 124, 133, 135, 138, 152, 154, 155, 184, 188, 202

age, 27, 118–20

alpha (α), 151

beta (β), 151

delta (δ), 151

discovery, 11–12, 54, 90, 155

epsilon (ε), 28, 151

eta (η), 94

gamma (γ), 151

structure, 90–2

UVIS. *See* Cassini spacecraft, Ultraviolet Imaging Spectrograph (UVIS)

Van Allen belts, 147
Van Allen, James, 147
vertical resonance. *See* resonance
viscosity, 37, 48, 50, 52, 53, 73, 77, 86, 108, 110, 117
viscous torque. *See* torque
vortices, 9
Voyager space missions, 2, 3–4, 6, 10, 11, 12–15, 16,
    17, 18, 26, 27, 28, 29, 30, 45, 58, 60, 66, 69,
    70, 71, 73, 77, 83–90, 91, 93, 94–6, 97, 98,
    109, 118, 120, 125, 126, 135, 136, 137, 138,
    139, 141, 142, 144, 149, 151, 152, 154, 157,
    158, 159, 160, 161–3, 164, 165, 166–9, 172–3,
    177, 178, 179, 182, 188, 189, 194, 196, 200,
    201, 202, 206

wakes, 4, 15, 17, 18, 29, 65, 66, 73, 76, 77–83, 104,
    106–7, 110, 150, 182, 204, 215, 216.
    See also self-gravity wakes
wandering shepherd. *See* shepherd
Ward, William, 15
Wasserman, Larry, 11, 12
water ice, 2, 11, 17, 25, 44–7, 129, 184, 192,
    200
waves, 2, 3, 4, 15, 29, 56, 58, 62, 63, 66–8, 73, 75,
    76, 85–7, 128, 187, 199, 204, 206, 211,
    212, 219. *See also* density waves
Weidenschilling, Stuart, 122
West, Robert, 135
Wisdom, Jack, 84

Printed in the United States
By Bookmasters